Neolithic cave burials

Manchester University Press

Social Archaeology and Material Worlds

Series editors
Joshua Pollard and Duncan Sayer

Social Archaeology and Material Worlds aims to forefront dynamic and cutting-edge social approaches to archaeology. It brings together volumes about past people, social and material relations and landscape as explored through an archaeological lens. Topics covered may include memory, performance, identity, gender, life course, communities, materiality, landscape and archaeological politics and ethnography. The temporal scope runs from prehistory to the recent past, while the series' geographical scope is global. Books in this series bring innovative, interpretive approaches to important social questions within archaeology. Interdisciplinary methods which use up-to-date science, history or both, in combination with good theoretical insight, are encouraged. The series aims to publish research monographs and well-focused edited volumes that explore dynamic and complex questions, the why, how and who of archaeological research.

Forthcoming

The Irish tower house: Society, economy and environment, c. 1300-1650
Victoria L. McAlister

An archaeology of lunacy: Managing madness in early nineteenth-century asylums
Katherine Fennelly

Images in the making: Art, process, archaeology
Ing-Marie Back Danielsson and Andrew Meirion Jones (eds)

Communities and knowledge production in archaeology
Julia Roberts, Kathleen Sheppard, Jonathan Trigg and Ulf Hansson (eds)

Early Anglo-Saxon cemeteries: Kinship, Community and Mortuary Space
Duncan Sayer

Urban Zooarchaeology
James Morris

An archaeology of innovation: Approaching social and technological change in human society
Catherine J. Frieman

Neolithic cave burials

Agency, structure and environment

Rick Peterson

Manchester University Press

Copyright © Rick Peterson 2019

The right of Rick Peterson to be identified as the author of this work has been asserted by him in accordance with the Copyright, Designs and Patents Act 1988.

Published by Manchester University Press
Altrincham Street, Manchester M1 7JA
www.manchesteruniversitypress.co.uk

British Library Cataloguing-in-Publication Data
A catalogue record for this book is available from the British Library

ISBN 978 1 5261 1886 8 hardback

First published 2019

The publisher has no responsibility for the persistence or accuracy of URLs for any external or third-party internet websites referred to in this book, and does not guarantee that any content on such websites is, or will remain, accurate or appropriate.

Typeset
by Deanta Global Publishing Services

Contents

List of figures		vi
List of tables		xi
Acknowledgements		xii
1	The body in the cave	1
2	In praise of limestone	11
3	Gestures and positions	41
4	How do caves act?	69
5	Origins	95
6	Written on the body	125
7	Deep time	156
8	Temporality, structure and environment	184
Appendix 1		213
Appendix 2		225
References		233
Index		254

List of figures

1.1	Known caves in Ireland and Britain with radiocarbon-dated human remains from the Neolithic period.	8
2.1	Human remains with Neolithic dates from caves in Europe.	12
2.2	Evidence of long-term funerary processes through the creation of clusters of disarticulated human bone at Ajdovska Jama, Slovenia.	17
2.3	Excavated evidence for human burial practices in the upper chamber at Grotta Scaloria, Puglia.	20
2.4	Earliest Neolithic single graves from L'Abri Pendimoun, Alpes Maritimes.	23
2.5	One of the postulated sequences for multi-stage burial at the Late Neolithic cave site of Can-Pey, Pyrénées-Orientales.	26
2.6	Cave burials of various dates in the Abri des Autours, Namur.	32
2.7	Neolithic human remains from Annagh Cave, Limerick.	34
2.8	Caves where human remains were deposited between 5500 and 5000 BC, showing the significant number of such sites from the French and Italian Early Neolithic but also the marked absence of such sites from the Eastern Adriatic coast.	36
2.9	Cave burials in Europe dating to the fourth millennium BC.	37
2.10	Caves with human remains for each 500-year interval during the European Neolithic.	38
2.11	Caves with burial activity dating between 5000 and 4500 BC.	39

List of figures

3.1	Michael Wysocki's reconstruction of the burial events in Wayland's Smithy 1.	45
3.2	Bara concepts of the person.	49
3.3	A Wataita cranial display niche in Kajire rock shelter, Tsavo, recorded by Kusimba and colleagues (2005), which contained 308 crania.	51
3.4	Sequences of bodily decomposition as suggested by palaeotaphonomic research. Shaded elements are those which have labile articulations and would be expected to disarticulate early.	55
3.5	Frequency ranges for the recovery of skeletal elements scavenged by canids in the north-western United States based on 53 missing persons cases.	56
3.6	Plan of the recorded position of human remains at Jama-Bezdan, Hrgar.	60
3.7	Phreatic portion of Fairy Holes Cave, Whitewell, Lancashire.	61
3.8	A variety of vadose erosional processes are visible in the entrance chamber of Dunald Mill Hole, Lancashire. These include the down-cutting of the floor by the streamway, which now flows beneath the boulders in the foreground of the picture, and substantial roof collapse caused by aerial weathering.	62
3.9	A solutional vertical shaft into the chamber at Heaning Wood Bone Cave, Cumbria.	63
3.10	Granular tufa deposit forming at the back of the Cave Ha 3 rock shelter, North Yorkshire.	64
3.11	Bands of alluvial silts and clays in section in Temple Cave, Whitewell, Lancashire.	66
4.1	The material culture of class as expressed in choices of tableware for social dining in France.	72
4.2	Probable beaver lodge in the Mesolithic organic deposit within the former course of the river Eden at Stainton West, Cumbria.	74
4.3	Ankave bark and rattan eel trap.	76
4.4	Deposition of different classes of material culture and human remains at George Rock Shelter, Goldsland Wood, Vale of Glamorgan.	87

4.5	A material index of the passage of time in an African house wall, in this case, an Ilkisongo Maasai house, Kajaido District, southern Kenya.	88
4.6	The Portesham mirror (centre) and the material indices which link it to other Iron Age objects, including shield edge binding; sheet metal cauldrons; sword scabbards; horse bridles; mirrors and pottery.	91
5.1	Location map for the sites discussed in Chapter 5.	96
5.2	Relationships between the excavated contexts and dated human bone at Carding Mill Bay 1.	101
5.3	Plan of the excavated deposits at Raschoille Cave.	104
5.4	The excavated area at the An Corran rock shelter, showing the extent of contexts 36 and 31, the upper midden layers where human remains were found.	106
5.5	Modelled dates for midden deposition in caves in the Torbryan valley, Devon.	109
5.6	Radiocarbon-dated human remains from early cave burial sites compared with the modelled posterior density estimates for the start of the Neolithic in south-west England, South Wales and Yorkshire and Humberside.	112
5.7	George Rock Shelter under excavation in 2007. The very light-coloured tufa-rich layer 1002/1007, which is where the bulk of the human remains were probably originally deposited, is clearly visible in section. Above this is context 1004, which also contained human bone and prehistoric artefacts. Close to the rock wall, the fill of the modern disturbance can be seen as a much darker area in section.	114
5.8	Modelled dates for the start and end of burial activity at George Rock Shelter.	115
5.9	The view from the interior of Kinsey Cave across the area of Mattinson's excavations in the entrance and the probable area of Neolithic burial activity.	117
5.10	The recorded position of human bone from within Thaw Head Cave.	118
5.11	Sewell's Cave, showing the area of the rock shelter against the northern wall, where the human remains were deposited.	120

List of figures

6.1	Location map for the sites discussed in Chapter 6.	126
6.2	Section through the excavated deposits at Lesser Kelco Cave.	127
6.3	The excavated areas at Robin Hood's Cave and the location of the human frontal bone fragment 465 and the other cranial and vertebra fragments.	129
6.4	The surviving elements of the Robin Hood's Cave head as reconstructed by Powers and Campbell.	130
6.5	Plan and section of the excavated deposits at Chelm's Combe.	136
6.6	Surviving elements in the Chelm's Combe skeletal assemblages from (left) the rock-cut chamber and (right) the granular tufa deposit in the rock shelter showing the different taphonomic signatures for each area.	137
6.7	View along the east fissure of Jubilee Cave towards the rock ledge where the burial was discovered.	140
6.8	The reconstructed location of the Little Orme Quarry human remains.	141
6.9	Plan and section of the excavated deposits within the Bower Farm rock shelter, showing the positions of the excavated human remains.	143
6.10	The interior of Cave Ha 3, showing the area of the hearth and the niches in the rear wall of the shelter.	144
6.11	Plan and section of the excavated area at Hay Wood Cave showing the find locations for the human crania and cranial fragments.	148
6.12	Section through the deposits outside Picken's Hole, showing the layer of matrix-supported scree (layer 3) where the human teeth were discovered.	153
7.1	Location map for the sites discussed in Chapter 7.	157
7.2	Plan of the caves at Inchnadamph showing the location of (1) the human cranium associated with disarticulated vertebra and sacrum and (2) the radiocarbon-dated human remains from the fissure at the back of the first chamber.	158
7.3	Plan and section of the excavated deposits in Backwell Cave.	160
7.4	Section through the deposits in Nanna's Cave.	162

7.5	Plan of excavated area P at Scabba Wood Rock Shelter, showing the location of the human remains and Neolithic worked stone.	164
7.6	Plan of the interior of Totty Pot and the location of some of the archaeological material.	167
7.7	Sketch section of the entrance shaft and rift at North End Pot, showing the approximate position of the human remains.	170
7.8	Simplified section through Ashberry Windypits I and II, showing the location of the Late Neolithic/Early Bronze Age human remains in chamber D.	175
7.9	Sections through the excavated parts of Charterhouse Warren Farm Swallet.	177
8.1	Duration of the 'head cult' burial rite in the Yorkshire and Derbyshire Pennines.	186
8.2	Probable duration of secondary burial rites in South Wales and south-western England.	187
8.3	Dated examples of secondary burial from caves and chambered cairns in South Wales and south-west England.	190
8.4	Modelled radiocarbon results from caves where there is good evidence for a successive inhumation rite.	192–193
8.5	The direction of aspect from the mouth of forty-two of the forty-eight caves in the study (dolines and other vertical entrance caves have been excluded).	199
8.6	Aspect directions of caves in the Yorkshire Pennines with Neolithic human remains.	200
8.7	View from the mouth of Kinsey Cave, facing south-west away from the high fells of Pen-y-Ghent and Ingleborough and across Giggleswick Common towards the Forest of Bowland.	201
8.8	The location of George Rock Shelter and of documented Early Neolithic archaeology on the Vale of Glamorgan.	205
8.9	The location of George Rock Shelter and the Early Neolithic archaeology in its immediate environs.	206
8.10	The location of Giggleswick Scar and of documented Neolithic archaeology in the surrounding landscape.	209
8.11	The location of Giggleswick Scar and the possible Neolithic archaeology in its immediate environs.	209

List of tables

3.1 Time sequences for the canid-assisted disarticulation of human remains. Based on 37 examples from the north-western United States.   57
5.1 Radiocarbon determinations from cave midden deposits in the Torbryan valley.   108
5.2 Human $\delta^{15}N$ (‰) values for dated cave burials in North Yorkshire.   122

Acknowledgements

This book was written over a considerable period of time. It had its origins in fieldwork carried out by myself and Stephen Aldhouse-Green between 2005 and 2007, and research I did then looking at comparative material for the Neolithic human remains we discovered at Goldsland Wood. I began to put the material into its present form from 2013 onwards, but over the whole extended period of thinking about Neolithic cave burial, I have incurred many debts.

The first of these are to the series editors, Joshua Pollard and Duncan Sayer, for their enthusiasm for incorporating the book into the Social Archaeology and Material Worlds series. At Manchester University Press, the book has benefitted enormously from the support, encouragement and patience of the commissioning editor Meredith Carroll and the hard work of Alun Richards. In addition to the valuable comments on the text at various stages by the series editors and anonymous reviewers, I have been greatly helped by the comments on various draft chapters which were kindly read by Vicki Cummings, Lindsey Büster, Andrew Chamberlain, Patrick Randolph-Quinney, Julia Roberts, Julian Thomas and Mick Wysocki.

The following people generously provided information and advice during the writing of the book: Olaf Bayer, Fraser Brown, Adrian Chadwick, Andrew Chamberlain, Vicki Cummings, Marion Dowd, Seren Griffiths, Chaprukha Kusimba, Luc Laporte, Jim Morris, Duncan Sayer, Rick Schulting, Peter Style, Patrick Randolph-Quinney, David Robinson, Sam Walsh and Mick Wysocki. Particular thanks are due to Seren Griffiths for her advice and discussion of the modelled chronology of the earliest cave burials in Chapter 5 and to Fraser Brown and Chaprukha Kusimba for permission to use the images reproduced here as Figures 4.2 and 3.3, respectively.

Archaeological fieldwork at Goldsland Wood was carried out through the kind permission of the landowner, Mr David Randolph,

Acknowledgements

and was greatly helped by the extensive support of the neighbouring landowners, John and Robert Reader. It was financially supported by the University of Central Lancashire, the British Academy, the Society of Antiquaries of London and the Robert Kiln Trust. The excavations were jointly directed by myself and the late Professor Stephen Aldhouse-Green with the assistance of Anne Leaver and Sam Williams. Radiocarbon dating on the George Rock Shelter human remains was carried out by Tom Higham at the Oxford Radiocarbon Accelerator Unit and supported by the NERC radiocarbon fund and the University of Central Lancashire. I owe a particular debt to Stephen, who first introduced me to cave research and was a highly supportive mentor during two important periods of my career, and who sadly died during the writing of this book.

The research and writing for this book was supported by two periods of university-funded sabbatical leave. I am very grateful for the support of my then heads of school, Gary Bond and Carol Cox, in being awarded these periods of leave. In any small academic department, sabbatical leave such as this also places a considerable burden on other colleagues. I have been very lucky to work as part of an extremely supportive team in Archaeology at the University of Central Lancashire. The hard work of Vicki Cummings, Seren Griffiths, Jim Morris, David Robinson and Duncan Sayer has contributed immeasurably to creating the space to write this book. Lastly, I need to acknowledge the tremendous and unstinting support offered throughout the period by my family: Julia, Ellis and Josie.

1

The body in the cave

During the Neolithic period in Europe, caves and other underground spaces were used for burial. The evidence for this practice is reasonably well understood but, with certain exceptions such as the Belgian Middle and Late Neolithic (Cauwe 2004), cave burial has usually been regarded as something tangential to the broader narrative of the European Neolithic. Caves are often treated as places for simple expedient burial, perhaps for less socially favoured members of society, when compared to an assumed norm of burial in monuments (see, for example, Schulting and Richards 2002a, 1021). In this book, I will discuss the human remains from British Neolithic caves on their own terms. They were part of a wider European tradition of cave burial. They were also an important strand in the overall diversity of funerary practice in the British Neolithic. By understanding cave burial in the period, we get a much clearer understanding of attitudes to death in all contexts.

One way of describing this book would be to say that it is an exploration of the archaeology and agency of natural places. However, it could also be described as a book about burial in British caves during the Neolithic period. Both of these descriptions are apt, but they reflect different traditions of research in archaeology. Research may be generalising and thematic and address globally applicable topics of past human existence – in this case, the archaeology of natural places and of human and environmental agency. Or it may be a particularising, locally situated investigation of the remains of a particular past time and place – in this case, cave burial in the British Neolithic. Both of these research traditions are important parts of how archaeology works. In this case, I hope that I have integrated general and particular research in a coherent way. This is not a book about agency with a case study about cave burial, nor a catalogue of cave burials with an interpretive conclusion based on actor-network theory. This is a discussion of some different conceptions of agency which I feel are particularly relevant and useful in trying to interpret the archaeology of Neolithic cave burial in Britain. It is not a complete review of the many different archaeological and anthropological uses of the term *agency*, and it certainly does not contain archaeological evidence from every known cave with Neolithic activity.

Of course, the idea that archaeology has something to say about natural places and the idea that animals, places and objects can be thought of as agents are not new. There is an extensive discussion of both of these topics from a range of different perspectives (e.g. Bradley 2000; Ingold 2000; Latour 2005). Similarly, Neolithic human remains from British caves have been reviewed by a number of writers (e.g. Barnatt and Edmonds 2002; Chamberlain 1996; Leach 2006; Schulting 2007). In excavating and researching Neolithic human remains from caves, I have consistently found myself addressing two problems which have provided a link between the general themes of agency and natural places and particular bodies in particular caves.

Neolithic burial and cave burial

The first of these questions is the problem of the relationship between these burials and other practices around human remains in the British Neolithic. Collective disarticulated burial in monuments is a particularly well-studied aspect of Neolithic studies (see, e.g., discussions in Wysocki and Whittle 2000 and Whittle et al. 2007), and two main interpretations of the burial process have been offered. Disarticulation may have been achieved through a multi-stage rite which involved some significant circulation of human bone away from burial monuments. This is often referred to in the literature as *secondary burial*; see Chapter 3 for a fuller discussion. Alternatively, the disarticulated state of bodies may be largely the result of taphonomic processes following the successive inhumation of bodies at burial sites. I will return to the details of this debate in Chapter 3, but it is clear that very similar arguments can be made about human remains from caves. It should also be borne in mind that Neolithic burial is not confined to cairns, long barrows and caves. From the Middle Neolithic onwards, there is a well-defined tradition of single burial, often associated with large round barrows (e.g. Gibson and Bayliss 2010, 101). Schulting (2007) has also pointed out the diversity of non-monumental burial. Human remains recovered from caves are usually discovered in an extremely fragmentary state. A very careful examination of the possible taphonomic processes is needed before we can draw parallels between cave burial and the range of other documented burial rites in the British Neolithic.

Do caves have agency?

The second question has arisen from a consideration of the nature of caves as spaces. Both Barnatt and Edmonds (2002) and, for Irish caves, Dowd (2008) have discussed the similarities of caves and chambered cairns as spaces. This provides another link between the monuments and caves beyond any possible similarities of burial rite. It is assumed that both caves and chambered tombs would have been thought of as conceptually similar places because they shared an architecture of passages and chambers. Barnatt and Edmonds (2002, 127) suggest that the practice of separating geological and architectural spaces into contrasting classes of natural and cultural entities is itself modern. This is a distinction which we cannot assume was made in the Neolithic, although, it should be noted that, more recently, Dowd (2015, 110) has suggested that caves and monuments *were* perceived as different to one another during the Irish Neolithic. She points to the different ways that human remains were disposed of in the two types of sites as evidence that they were not perceived as equivalent spaces.

More broadly, Barnatt and Edmonds (2002, 125–127) and Dowd (2008, 311–312) both provide a wider consideration of the phenomenological impact of these constricted spaces. This is an area which has been particularly explored by European scholars: for example, in Mlekuž's (2011) work on the Italian and Slovenian karst. Mlekuž studies the impact of physically inhabiting caves and rock shelters on the bodies of both sheep and shepherds. The cave walls cease to be something which is merely a passive arena within which human and animal actions take place. The walls themselves 'push back'. In a similar vein, Bjerck (2012) has examined how darkness and constriction influenced the placing of Bronze Age rock art in Norwegian caves. These discussions of the power of cave spaces to act on people lead us to a wider debate about whether inanimate objects like caves have agency. It is clear that caves *can* do things to people; the question is really about whether it is enlightening or convincing to describe this effect as *agency*. I have explored this debate about the agency of caves previously in relation to later prehistory (Peterson 2018). In that work, I argued that caves would have been understood in the past as possessing agency and that it is helpful to think of them in these terms. However, we also need to be aware of the dangers of treating agency unreflectively. If we reify 'agency' as a social force to the point where it becomes the explanation, then the idea ceases to have any value as a conceptual tool. For this reason, I have suggested, in this book and in the aforementioned work, that we re-phrase the question around cave agency to ask 'how did caves act on people?'

Therefore, this book will attempt to tackle two problems. Firstly, how do cave burials relate to other Neolithic burials? Secondly, how do caves act on people? These two questions belong together because of the way that burial practice links society and environment. If we return to the division of burial practices into either secondary burial rites or successive inhumation, then one of the ways of distinguishing between them is to look at the agent of disarticulation. A secondary burial rite involves repeated interventions from living people. Bodies must be laid out and transported, and often they are physically broken up. Bones must be recovered, sorted and ultimately placed in a final burial site. Through all of these processes, the agency of living humans – the mourners or descendants – is the main driver of the physical process of disarticulation. By contrast, when bodies are placed successively in either a tomb or a cave, then the main agent of disarticulation is a combination of time, the physiological properties of the decaying corpse and the physical properties of the space of burial. This is not to suggest that time and environment are not important in many multi-stage rites, or that people could not interact with successively inhumed bodies during decomposition if they wished. However, human agency is necessary for secondary burial rites, and natural agency is an essential part of successive inhumation. Thinking about the relative contributions of society and the environment to the burial process gives us a common thread to our answers to both of the problems I posed at the start of this paragraph. Cummings (2017, 94) has argued that the 'normal' fate of human remains in the Neolithic was a rite of transformation primarily driven by natural agents such as scavenging animals and bodily decomposition processes. She postulates that most bodies were exposed and scavenged to the point where they were completely broken down and destroyed. From this perspective, what is distinctive about secondary burial or successive inhumation, whether it took place in a cave or a monument, is that it removed a body from this complete transformation and allowed some traces of it to survive. Within Chapter 3, I examine not only the anthropological evidence for the social customs and structures which may have surrounded secondary burial and successive inhumation but also the detail of the processes of bodily decay and cave sedimentation which would have been the natural agents of change. In Chapter 4, I have tried to further draw out the implications of treating inanimate objects as having agency. Caves, material culture, bodies and time are all considered from the standpoint that it is unhelpful to maintain a strict division between living subjects and inanimate objects.

Dated Neolithic human remains from British caves

If we want to analyse burial practices in caves in the Neolithic, our first requirement is data: a selection of cave sites where we know human remains were deliberately deposited during the period. There are many cave sites where Neolithic artefacts have been found alongside human remains; for example, Barnatt and Edmonds (2002, Table 1) list twenty-five such sites from Derbyshire alone. However, the analysis of Neolithic cave burial practice would not be possible without the radiocarbon dates on human bone provided by many different research projects over the past 20 years. These dates are absolutely essential. Previous studies of caves and human bone taphonomy, particularly by Leach (2006, 2008), have shown that radiocarbon dating is the only reliable guide to determining the date of a cave burial. Conventional archaeological assumptions about the integrity of sealed contexts and associations between artefacts and human bone cannot always be relied upon in cave environments. The open texture of many scree deposits and the highly active geological processes within cave systems mean that it is extremely common for artefacts and human bone to be moved, re-deposited and combined in complex ways.

Some of the burials I discuss can be used as examples to reinforce this point. As has been previously noted (Schulting 2007, 586), many of the bones were originally sampled as part of projects investigating the Palaeolithic use of caves. They were submitted for dating because they were thought to be securely stratified in Pleistocene contexts. For example, the burials from Cattedown Cave in Devon have Neolithic dates but were discovered in a breccia deposit beneath a stalagmitic floor (Worth 1887, 110), and they were dated on the understandable assumption that both the breccia and the flowstone above it were *in situ* Pleistocene deposits (Higham et al. 2007, S28–S29). Therefore, if we are to study Neolithic cave burial, only those sites with direct dates on human bone should be considered. While this undoubtedly excludes some caves which were used in the period, a clear comparison with the European data, with burial in monuments and with other cave burials requires the use of absolute dating.

Forty-eight directly dated Neolithic cave sites in Great Britain have been used in this study (see Appendix 1 for the complete list). All of these sites have at least one published radiocarbon date on human bone which, when calibrated to two standard deviations, falls into the Neolithic period. For the purposes of this book, I have taken the view that any date which has part of its calibrated range between 4000 and 2400 BC should be included in the table. There are a further nine sites where Neolithic radiocarbon dates were obtained from the Oxford

AMS facility but which were subject to problems caused by ultrafiltration contamination (Bronk Ramsey et al. 2004; Rick Schulting, personal communication). These sites are Carsington Pasture Cave and Fox Hole Cave, both in Derbyshire; Gop Cave, Flintshire; Happaway Cave, Devon; Ifton Quarry Rock Shelter, Monmouthshire; Ogof Pant-y-Wennol, Llandudno; Red Fescue Hole and Pitton Cliff Cave, both on Gower; and Priory Farm Cave, Pembrokeshire. I have given a full list of these sites here, as some have already featured in published discussions of Neolithic cave burial (e.g. Barnatt and Edmonds 2002, 114–116). Indeed, it is highly likely that burial took place at most of these sites during the Neolithic. However, in view of the problematic nature of the dates, they were not included in this study. During the final revisions of the text of this book new dates became available as a result of ongoing aDNA studies (Brace et al., in preparation) which confirmed an Early Neolithic date for Carsington Pasture Cave and identified further sites with directly dated Neolithic human remains at Aveline's Hole and Ogof-yr-Ychan.

The time range of 4000–2400 BC for this book has been chosen to ensure that the study covers the processes around the beginning of the Neolithic. Andrew Chamberlain was the first to point out (1996, Figure 1) that there was a substantial increase in the deposition of human bone in caves around 4000 BC. This data has subsequently been refined by Schulting (2007, Figure 2), and both authors agree that there is evidence for a significant new practice of cave burial in the centuries around 4000 BC. This is interesting, as it means that new cave burial practices were being introduced at approximately the same time as farming, substantial buildings, monuments, polished stone tools and pottery, all traits which we identify as part of the beginning of the Neolithic. However, thanks to the large-scale use of Bayesian statistics on radiocarbon data sets, we now have the beginnings of a much more precise chronology for the adoption of the Neolithic in different regions (Griffiths 2014a, b; Whitehouse et al. 2013; Whittle et al. 2011). This means that the exact relationship between the beginnings of cave burial and the adoption of the Neolithic needs to be addressed. Some of the early fourth-millennium cave burials in Britain could potentially have been Late Mesolithic rather than Early Neolithic, especially in the north and west. This is a point which has been debated previously. Hellewell and Milner (2011) consider that a Mesolithic date could be established for at least some cave burials, and they propose that cave burial was an example of continuity between the Late Mesolithic and the Early Neolithic. Schulting and colleagues (2013, 22) come to exactly the opposite conclusion. They point to the significant increase in burials dating to the Early Neolithic as evidence of an independent development of cave burial at this time. This is a highly complex area which requires a clear distinction to be made between 'the Neolithic' as a chronological marker and 'the Neolithic' as a description of a way of life. This debate forms the core of Chapter 5, which examines the likely origins and date of the first fourth-millennium BC cave burials in Britain.

It is also necessary to be cautious when using the radiocarbon dates from caves to discuss the likely duration of burial activity, either at individual caves or in the Neolithic period generally. The same problems of open scree deposits and active cave processes mean that we can very rarely prove that two dated bones from the same layer come from a single phase of activity, or that dated bones from superimposed layers represent a sequence of burial events. Despite these limitations, we do have good data that allows us to demonstrate that there was a range of different Neolithic burial practices in caves. The stratigraphic problems do not stop us from analysing and comparing these different practices and attempting to answer the linked questions about burial practice and the agency of caves. Additionally, it *is* possible to use Bayesian methods to discuss the likely chronology of these different practices between sites and across regions. These detailed chronologies are discussed most fully in Chapter 8.

Figure 1.1 indicates the distribution of the radiocarbon-dated Neolithic human remains from caves in Britain and Ireland used in this study. This distribution is at least partly influenced by the availability of suitable caves for burial. The published data (Chamberlain 2014) on caves which contain human remains of any date can be used as a proxy to show which caves would have been available for burial in the Neolithic. On this basis, in the southern part of the country, we can see that wherever there were groups of suitable caves, then there was Neolithic cave burial. To the north of Yorkshire, however, there are large areas which have suitable caves without any Neolithic burials. This is not solely a result of where fieldwork is being carried out. For example, the group of caves along the south coast of Cumbria has been the subject of a recent research project (Smith 2012a) which included radiocarbon dating on human bone, but no Neolithic burials have yet been identified in this area.

Burial and time

We have the potential to understand the chronology of human remains in caves from this period. It could be argued, however, that this is only part of the answer. Chronology, as measured by radiocarbon dating, is not necessarily the same thing as the human experience of

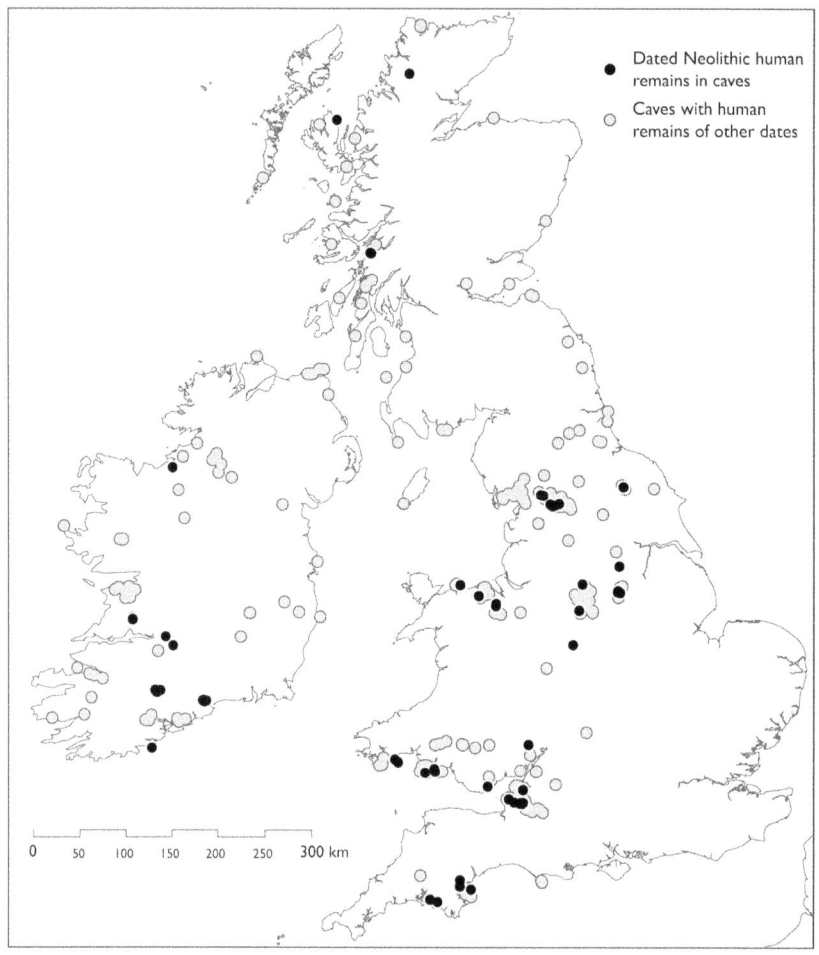

1.1 Known caves in Ireland and Britain with radiocarbon-dated human remains from the Neolithic period (based on data from appendix 1 of this volume; Dowd 2015, table 5.1 and Fibiger 2016). The grey circles show the positions of caves in both countries with human remains of any date (British data from Chamberlain 2014, Irish data from Dowd 2015, Appendix 1). Base mapping of Ireland © Ordnance Survey Ireland and Ordnance Survey of Northern Ireland provided under creative commons 4.0 international licence. Base mapping of Great Britain © Crown Copyright and Database Right (2017). Ordnance Survey (Digimap Licence).

the passage of time. To understand the way in which caves, bodies and people interacted around cave burial, we need to have a similarly embodied and experiential view of the way in which they experienced the passage of time. There is an extensive archaeological and anthropological literature on the human experience of time which is relevant here. Of particular importance is Tim Ingold's (1993, 159) discussion of temporality. In this work, he coined the neologism 'taskscape' to help discuss time in an embodied and experiential way. If a landscape is thought of as an array of geographical features, then a *taskscape* is an array of activities; in both cases, the arrays are connected by being experienced by a participant. At the heart of the taskscape is an understanding of time derived from phenomenology, particularly from Merleau-Ponty (1962, 416–421), which depends on human experience rather than any external constant. This temporality derives directly from actions: when people do things, they make time pass. This argument will be developed in more detail in Chapter 4, but temporality is key to understanding both the processes of burial and the actions of people and caves. Past traces of earlier activities would have provided the structure for these burials to take place. Decay processes and geomorphological change in caves would have given material indications of the passage of time. The actions of caves, dead bodies, artefacts and people would have been understood in terms of the passage of time. A key distinction between the different kinds of burial rites, both within caves and elsewhere, would have been their different temporalities.

In Chapters 6 and 7, I will argue that we can see a major change in cave burial practice during the Neolithic period. Early Neolithic cave burial in particular was very diverse and is best interpreted by comparison with a whole range of wider contemporary traditions about human remains. For example, midden burials from caves form part of a wider tradition of midden burial, particularly in western Scotland, during the fourth and fifth millennia BC (Milner and Craig 2009). The practice of successive inhumation in the Early Neolithic may have been similar, whether it was taking place in a cave or in a chambered cairn (Leach 2008, 46–48; Wysocki and Whittle 2000, 595–598). At this date, we seem to have evidence for a range of different rites in caves, each more similar to a different non-cave rite than to other types of Early Neolithic cave burial. However, later in the Neolithic period, it seems as if the range of practices associated with caves had become much more restricted. I will argue that by around 3200 BC, there was a genuinely distinctive cave burial tradition which was coherent and noticeably different from non-cave burial practices. Later period burials like this were generally deeper into the cave, had less opportunity for living people to be involved in the processes of decomposition and disarticulation and drew more strongly on the particular affective and geomorphological properties of caves. The interaction between the agency of two natural processes – bodily decomposition and cave geomorphology – and the social agency of the mourners carrying out the rites seem to have led to the development of a style of burial specific to caves. Some aspects of this rite continued into Beaker and Early Bronze Age period cave burials. However, as might be expected in a period with a distinctive and well-understood set of funerary rites for non-cave burials, Early Bronze Age cave burial seems to have its own different set of rites and practices. The details of these, unfortunately, take us beyond the scope of this book.

In the following chapters, I will attempt to set out these arguments in more detail, to describe the variety of these different burial styles and to offer an account of the broader principles behind the development of a recognisable Neolithic cave burial rite. The common factor in all of these rites was their long duration. Almost all Neolithic burial seems to have been an extended process, presumably aimed at providing a managed transition through the social complexities of mourning and the physiological processes of decay. Extended burial rites like this have been widely studied in a range of disciplines. In Chapter 3, I shall provide a review of these interpretations and the history of their application to Neolithic burial.

The body in the cave, therefore, can be understood as a central part of the British Neolithic. However, British cave burial is only a subset of a wider European phenomenon of cave use in prehistory. Trying to understand the reasons for the adoption of cave burial in Britain around 4000 BC clearly depends on an understanding of both the longer time depth and wider archaeological context available from the continent-wide evidence for similar practices. Wherever there are suitable caves throughout Europe, there are human remains which can be shown to have been deposited during the Neolithic period. The overall spread of the European Neolithic from its Near Eastern origins is reflected in the date and distribution of these cave sites. The earliest examples are in the eastern Mediterranean, with the British caves forming part of the relatively late group in western and northern Europe. Chapter 2 is concerned with this European context in more detail. I have tried to unpick the evidence for different cave practices around human remains in all of the regions of Europe. This will provide a robust background and context to the description and interpretation of the changing practices in British caves in the rest of the book.

2

In praise of limestone

The deposition of significant numbers of human remains in British caves appears to have started around the beginning of the Neolithic period. This was, of course, late in the overall European Neolithic sequence. Therefore, cave burials from across Europe have the potential to help us understand this process. Did cave burials occur in Britain after 4000 BC because cave burial became more common everywhere at this date? Was there a 'European cave burial horizon' which just happened to broadly coincide with the start of the British and Irish Neolithic? Alternatively, was cave burial one of the group of associated new practices, such as farming, settlement and the use of pottery, which we now recognise as 'Neolithic'? Was cave burial, in all its complex variations, one of the ideas which spread as part of the European Neolithic and was it therefore introduced to Ireland and Britain as part of the process of becoming Neolithic?

Any book which is concerned with the archaeology of caves is also, by definition, going to be focussed primarily on limestone landscapes. The title of this chapter comes from W.H. Auden's poem of the same name, in which, among other things, he celebrated the mutable and active nature of limestone landscapes. In a critical essay reprinted in the collection *The Dyer's Hand*, Auden described the landscape characteristics of his personal Eden in the following way.

'Limestone uplands like Pennines plus a small region of igneous rocks with at least one extinct volcano. A precipitous and indented sea-coast.' (Auden 1962)

Auden's – slightly tongue-in-cheek – vision provides us with a precis of the kind of environment, in Greece, Italy and the Balkans, which was important in the creation of the European Neolithic. For Auden, the qualities he ascribed to limestone provided a unifying narrative to link an idealised southern Europe with his native Yorkshire. In a somewhat similar manner, I want to argue that the archaeology of Neolithic cave burial connects regions which are as apparently different to one another as the Peak District, the Meuse basin, Provence and Puglia. In this chapter, I will review some of the evidence for Neolithic cave burial rites in the limestone regions of Europe. I will also look at how these cave burials fit within their local Neolithic sequence. In some regions, cave burial is an important strand of evidence for the earliest Neolithic, and it is therefore plausible to argue that its adoption is connected with the changing practices and worldviews associated with becoming Neolithic. In other areas, however, the large-scale use of caves for burial is a Middle or Late Neolithic phenomenon. In this case, it is more plausible to argue that cave burial relates to different processes.

Societies in transition: The Neolithics of Europe

The date and character of the Neolithic in Europe varies. However, for the purposes of this book, I am particularly interested in how the Neolithic spread to those areas of Europe where caves were used for burial and what kind of Neolithic was present in those areas. Cave burial in Europe was not uniformly distributed (see Figure 2.1) and,

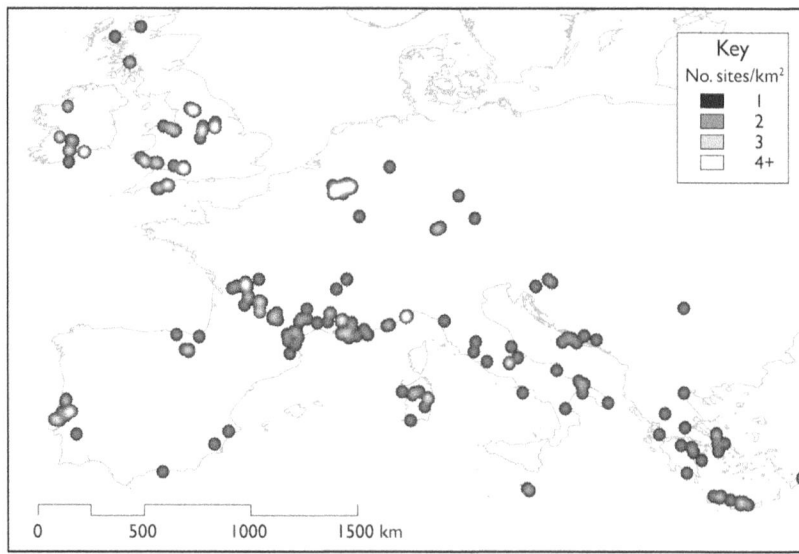

2.1 Human remains with Neolithic dates from caves in Europe. The data are presented here as a heatmap, which, at this scale, shows the density of activity within some particular small areas, such as the Meuse Basin or the Yorkshire Pennines, more effectively than a simple distribution map. See Appendix 2 for the original sources for this data (total number of sites = 262). The base mapping includes data licenced from © EuroGeographics.

alongside the obvious constraints of geology, the data in this chapter demonstrates that cave burial was more widely used in some areas and at some times than others. Therefore, the overall context for the practice of cave burial was provided by different regional variants of the Neolithic.

At the broadest scale, Neolithic practices, particularly farming, sedentism and the use of pottery, can be seen to have spread into western Europe by two routes. These are associated with the Linear Pottery Culture (Linearbandkeramic [LBK]) identified in Central Europe (see e.g. the reviews by Gronenborn 2007; Gronenborn and Dolukhanov 2015; Hofmann 2015); and with the Cardial Impressed Ware complex present along the western Mediterranean region (Guilaine 2015; Guilaine and Manen 2007). The earliest cave burials in western Europe occurred around the Mediterranean. It was not until the fourth millennium BC that we can see significant evidence for human remains from caves in north-western Europe. The precise process of change remains an area of debate in almost all areas of Europe, with different levels of emphasis given to the role of migrating populations and local innovation (e.g. Binder and Maggi 2001; Cassen 1993; Mlekuž et al. 2008). The availability of aDNA results for some key areas of Europe has led to a renewed focus on population movement as a mechanism for the transition (e.g. González-Fortes et al. 2017; Haak et al. 2010, 8–10). However, archaeological evidence from western regions of Europe has often been interpreted as a mosaic of local adaptations responding to the introduction of these new sets of knowledge (e.g. Cummings and Harris 2011; Louwe Kooijmans 2007; Vanmontfort 2008). Ancient DNA evidence from western regions of Europe has also shown, in some cases, that population movement was not a significant contributor to change (Jones et al. 2017). We can therefore imagine the Early Neolithic communities of the western limestone regions of Europe as immigrant farmers, transformed hunter-gatherers or, probably more realistically, as complex hybrid societies created out of migration, inter-marriage, raids and feuds, gift exchange and emulation between neighbouring groups. As Robb (2013) has pointed out, the significant thing about all these possible local pathways to the Neolithic is that, once a Neolithic way of life had been adopted, it was difficult for societies to revert to hunting and gathering. Robb (2013, 665–670) has demonstrated how, over the whole continent, the material consequences of the environmental and social processes of becoming Neolithic would have the effect of both making the transition to the Neolithic irreversible and of creating increasing convergence between the types of Neolithic society created.

Human remains from caves occur wherever there are caves in Europe. Large amounts of this evidence come from the seventh to the fourth millennia BC, when this transition to the Neolithic was taking place. In the following sections, I have attempted to review how Neolithic human remains from caves fit into the wider evidence for the Neolithic for each region. There is not the space in this book to attempt a critical synthesis of the European Neolithic as a whole. The most up-to-date examples in English are the papers in Fowler and colleagues (2015) and the volume by Whittle (1996). Chapters 2, 3 and 4 of Thomas (2013) also provide a thorough review of the European evidence as it relates to the spread of the Neolithic into Britain and Ireland.

Greece and the Eastern Balkans

The region of Europe with the earliest evidence for cave burial is Greece, which is unsurprising as it is also the region with the earliest Neolithic sites. There is some Early Neolithic burial evidence from Franchthi, Theopetra and Tsoungiza Caves (Tomkins 2009, Table 2) but despite this, human remains from caves are generally late in the Greek Neolithic sequence. The Greek earlier Neolithic was characterised by long-lived permanently occupied open settlements and arable production, and it has usually been assumed that it was directly derived from the first Near Eastern farming settlements (Demoule and Perlès 1993, 364–365). The broad outline of the Greek Neolithic presented here follows the chronology suggested in Perlès (2001) and Tomkins (2009). The Initial Neolithic period appears to have lasted from around 7000 BC to between 6500 BC and 6400 BC. The subsequent Early Neolithic period lasted until at least 6000 BC. During this phase, there is evidence for the relatively rapid introduction of agriculture, permanent settlement and pottery over Thessaly and Southern Greece. This is also the time when we see the first evidence for substantial houses in the region (Perlès 2001, 98–110).

Well-dated examples of human remains from Greek caves fall mostly into the Late or Final Neolithic (see Appendix 2), after around 5300 BC. Tomkins (2009 and 2013) has argued that these later Neolithic human remains are part of a wider, poorly recognised, set of evidence for the ritual importance of caves in the Greek Neolithic. He suggests (Tomkins 2013, 62–65) that the presence of occasional disarticulated fragments of human bone is one of the traits which mark out Neolithic caves in the Aegean as a different kind of place than the contemporary open-air settlements. Relatively low levels of fragmented human bone are part of a set of characteristics, including separation from areas of agricultural land, darkness, constriction and the fragmentation of material culture, which mark these cave sites as the precursors to the better known cult caves of the Cretan Bronze Age.

Tomkins (2013, 62) also points to a different tradition which led to the presence of larger quantities of human bone in later Neolithic caves in the region. There is evidence of successive inhumation, leading to the *in situ* excarnation of bodies, at most of the caves in the region which have been excavated in recent times (Tomkins 2009, Table 2). At some sites, for example, Genari Cave on Crete, articulated remains survived on the surface, and the large numbers of fragments from Late to Final Neolithic layers at Alepotypra and Kitsos Caves lead Tomkins (2009, 141–142) to identify a long-term process of excarnation at these sites. However, it is clear, from both the numbers of individuals involved and the relatively low numbers of caves with extensive collections of human remains, that this was a minority treatment for the dead.

In the later Neolithic, after about 5300 BC, there is evidence for a varied Neolithic in Greece and the Eastern Balkans. The population apparently increased, and more and different styles of settlement and many new settlement sites were created. The use of cave sites for all purposes also increased substantially in this phase (Tomkins 2009, 127). Importantly, there is some evidence to suggest differentiation in the types of farming that were going on in different regions, for example, nomadic pastoralism in central Macedonia (Demoule and Perlès 1993, 388–390). During the Greek Final Neolithic, after around 4500 BC, there is increasing evidence for these different local experiences of the Neolithic. In southern Greece and the Aegean islands, there were more small and dispersed settlements, and seasonal pastoralism seems to have become increasingly important. Further north, fewer sites were occupied than in the earlier phases, but the remaining settlements were large (Tomkins 2009, 127). This is particularly true in Macedonia, where early Final Neolithic settlements expanded in size (Demoule and Perlès 1993, 398–400). This pattern is also reflected in the northern parts of the eastern Balkans where, after about 4000 BC, large scale settlements were replaced by smaller complexes of short-lived pit huts. Bailey (2001, 259–261) suggests that this marked a fundamental shift in the way people drew upon the experience of living in houses and villages. He sees this as being replaced, at least partially, by a symbolic permanence based on grave-mound cemetery sites.

When we compare the evidence from Greek caves with interpretations of the wider Neolithic in the region, then we can see that cave burial cannot plausibly be interpreted as part of the process of becoming Neolithic. The significant periods appear to be the Late and Final Neolithic. It is at this point that some caves began to be used for collective burial, and human remains were deposited in others as part of the possible ritual use of caves. Drawing on Bailey's (2001, 259–261) insight, that cemetery sites may have functioned as markers of symbolic permanence in the northern Balkans, it is possible that human remains in caves performed the same function in the south. With increasing evidence for pastoralism and less permanent settlement, cave burials may have become important memorialised points in a seasonal pastoral round.

The Eastern Adriatic

In the western Balkans, along the east coast of the Adriatic, Neolithic human remains from caves are also relatively rare. This is particularly noticeable given the high numbers of natural caves in the region and indeed the much larger number of caves with Neolithic archaeology but no recorded human bone (Trimmis 2016). The regional Early Neolithic lasted from 6000 to 5500 BC (Forenbaher et al. 2013). Early in this period, evidence from caves shows that small groups of pastoralists were using pottery and herding domesticated sheep or goats. After about 5750 BC, permanent farming settlements developed in the region (Forenbaher and Miracle 2013, 72–74). Of more direct relevance to cave burial is the development of the Late Neolithic in the region. This is associated with the Hvar pottery style, which Forenbaher and colleagues (2013, 604) would see as lasting from 4800 to around 4000 BC. Mlekuž (2005) has carried out detailed analysis of the kinds of pastoralism associated with cave sites in the eastern Adriatic. He has demonstrated that the early pastoralists were highly carnivorous, killing animals from the herd for immediate consumption, but that by the Late Neolithic, the pastoral economy of the people using the caves had become more complex, with specialised roles for different species and the probable development of dairying (Mlekuž 2005, 42–43).

Forenbaher and colleagues (2013, 351–352), note five examples of human remains from caves in coastal regions of Croatia. Two more sites are known from Slovenia (Bonsall et al. 2007; Mlekuž et al. 2008) and, as with the Greek examples listed here, these are largely Late Neolithic in date (see Appendix 2). The site of Grapčeva Cave, Hvar, Croatia has been interpreted as a focus for mortuary ritual connected with the secondary burial of the remains of at least seven people (Forenbaher et al. 2010, 350–352). They suggest that selected skeletal elements were brought to the site after an intermediary period of burial elsewhere. They argue that the relative lack of similar sites in the eastern Adriatic points to a special status for Grapčeva as the focus of an underground cult, similar to those connected with the presence of 'abnormal water' identified in the Italian Neolithic by Whitehouse (2015, 57–58).

The Slovenian site of Ajdovska Jama (see Figure 2.2) provided slightly different evidence for a larger-scale mortuary ritual (Bonsall et al. 2007, 730–731) with the disarticulated remains of at least thirty-one people apparently exposed in the main chamber of the cave and then collected into discrete bone clusters towards the cave walls. Once again, this was a Late Neolithic practice: direct radiocarbon dating on human bone shows that this took place over a short period around 4300 BC (Bonsall et al. 2007, 734). At a similarly late date, small quantities of disarticulated and fragmented human bone were being deposited alongside pottery and animal bone at Mala Triglavca cave (Mlekuž 2012, 209; Mlekuž et al. 2008, Table 2).

2.2 Evidence of long-term funerary processes through the creation of clusters of disarticulated human bone at Ajdovska Jama, Slovenia (after Bonsall et al. 2007, Figure 3).

For the Balkan region considered as a whole, cave burial can be seen to be both relatively rare and relatively late in the local sequence. When the evidence from both Greece and the Eastern Adriatic is considered together, it suggests that there could be a connection between the deposition of human bone in caves and the development of specialised complex pastoral ways of life.

Italy

Cave burial in the Italian Neolithic marks a distinct change from the pattern seen in the Balkans. Caves were used as part of the normal repertoire of responses to death (Robb 2007, 56–58) from the beginning of the Early Neolithic. Robb (2007, 24–25) has reviewed the evidence for the beginning of the Neolithic in Italy. This seems to have begun first in the south-eastern region of Puglia around 6100 BC, with the full Early Neolithic beginning around 5800 BC (Skeates 2005, 18–19). By 6000 BC, there was some Neolithic activity in separate areas further north on the Adriatic Coast, across parts of central Italy and in Sicily. Neolithic activity in Italy spread further by 5500 BC to include the coastal regions facing the Ionian Sea and, in a separate development, the Ligurian Coast in the north-west of peninsular Italy (Robb 2007, 24–25).

Simple single inhumation pit burials were common during the whole of the Neolithic, with very similar style burials occurring on both settlement sites and in caves (Robb 1994, 36). Zemour (2011, 261) suggests, in a review of wider north-west Mediterranean Early Neolithic burial evidence, that the homogeneity of this practice may have been overstated and that a wide range of burial practices may have led to these apparently simple burials. Specifically discussing the Italian evidence, Robb (2007, 57–60) points to the widespread practice of burial disturbance as evidence that pit burial was just the first part of a multi-stage rite. The normal biography of a corpse, especially one buried in a cave, ended in the disturbance and the scattering of at least some of the bone. There were also other burial practices in Neolithic Italy which cross over between caves and open air sites. Robb (2007, 58–60) notes that there are both burials where the head appears to have been removed at some date after the burial took place, for example, at Cala Colombo cave cemetery, and apparent examples of the curation and display of crania. Human remains from some sites may be more directly related to wider ritual practices. Robb (2007, 60) gives the example of an adult male from Grotta Patrizi associated with a structured deposit of artefacts and animal bones.

The Middle Neolithic in Puglia (Skeates 2005, 18–19) began around 5700 BC with the transition to the Late Neolithic occurring at around 5400 BC. By 5000 BC, people living a Neolithic life occupied all most of lowland Italy and the offshore Islands of Sardinia, Corsica and Malta (Robb 2007, 24–25). The Late and Final Neolithic in Italy was marked by an increase in the importance of pastoralism. Robb (2007, 312–313) sees this as a specific cultural choice to intensify animal production over any other aspect of the farming regime. There was also a general trend for fewer, more dispersed settlements (Robb 2007, 303). At the same time, burial became much more prominent, with the development of specific cemetery sites of repeated cist burials (Robb 2007, 306). In a somewhat similar way to the evidence from the northern Balkans noted here, peoples' connection to their immediate environment was now marked in death, rather than in long-lived settlement structures.

There is debate about the connection between cave burial and the wider use of caves for ritual and cult purposes in the Italian Neolithic. Robb (1994, 36–37) maintains a clear separation between the two kinds of cave. He points to the distinction between the deposition of human remains and other cult activities at Grotta Scaloria, Puglia, and interprets most cave burial as part of a wider pattern of burial in the vicinity of settlement. There is some evidence for a distinctive set of practices around cave burial and disarticulation: 53% of burials from open-air sites and 88% of burials from cave sites became disarticulated (Robb 2007, Figure 9). The complexity of the interaction between cult use of caves and cave burial practices more generally is brought out in detail at Grotta Scaloria (Robb et al. 2015 and see Figure 2.3). There are two separate chambers at Grotta Scaloria; the lower set of passages is long and difficult to access, and it was this part of the cave system which was the focus of the cult of 'abnormal water' discussed by Whitehouse (2015, 57–58). There are human remains from this section of the cave. The disarticulated remains of one individual were discovered alongside the articulated remains of another buried in the sitting position (Robb 1994, 55). The upper chamber is larger and contains the remains of between twenty-two and thirty-one individuals and has evidence of at least five different burial rites (Robb et al. 2015, 41–42). Most of the bone from the upper chamber had been brought to the site as a combination of whole and partially fleshed bodies. These were then manually disarticulated and de-fleshed with stone tools before being discarded in a co-mingled layer with animal bone, stone tools and pottery. Robb and colleagues (2015, 49) interpret this as the final 'cleaning' event in a secondary burial rite. Other human bone from the upper chamber at

2.3 Excavated evidence for human burial practices in the upper chamber at Grotta Scaloria, Puglia (after Robb et al. 2015, Figure 2). The co-mingled human remains were most common in trench 10 but were present in all the excavated areas. The single adult cranium was found in trench 1, the juvenile burial in trench 6 (Robb et al. 2015, 41).

Grotta Scaloria included two pit burials. One of these was a juvenile of between 5 and 7 years old, from which the head had been removed at some point after burial. A single adult cranium was also found upright on a small stone niche. All these practices are likely to have taken place between 5500 and 5200 BC. There are also some single burials with grave goods with slightly later dates, towards the end of the sixth millennium BC (Robb et al. 2015, 42). Skeates (2013, 34) has interpreted the upper-chamber deposits at Grotta Scaloria as part of a wider practice in south-east Italy of cave-based mortuary feasting, which was accompanied by the conspicuous consumption and deposition of objects of value. Therefore, at Grotta Scaloria, we can see secondary burial rites, evidence for curation and manipulation of the head and single inhumation in pits in the upper cave. All of these were part of a broader Late Neolithic set of funerary traditions. In the lower cave, a distinctive and complex funerary practice was directly associated with the cult of 'abnormal water', which again had parallels with other caves in south-east Italy such as Grotta di Porto Badisco (Whitehouse 2015, 57–58).

The islands around the western and southern coasts of Italy show evidence of similar practices (see Appendix 2). Most caves of the Early Neolithic in Sardinia have been interpreted as habitation sites, although two have burial evidence. At one of these, Riparo sotta roccia Su Carroppu, two contracted burials were found at the back of a cave which contained extensive midden deposits. The other site, Grotta Verde, appears to be an early example of complex funerary behaviour associated with what Robb (2007, 60) would define as a ritual cave. Here, human remains and pottery had been deposited deep in the cave system on ledges around a subterranean freshwater lake (Skeates 2012, 168–170). The use of caves increased in the Middle and Late Neolithic, but human bone from these sites is still relatively rare. At Grotta Rifugio, disarticulated human remains, presumably representing the final phases of multi-stage burial rites, were discovered in the deepest part of the cave system. A minimum of twelve individuals were associated with several thousand perforated shells and other beads. Grutta I de Longu Fresu contains painted rock art and has been interpreted as a ritual cave. Here, too, disarticulated human bone appears to have been part of a multi-stage burial rite (Skeates 2012, 171–172). More human remains from the Final Neolithic and Earlier Copper Age have been discovered in caves in Sardinia, involving five sites in total. Some of these, such as Grotta Filiestru, have small quantities of human bone within large and long-established midden deposits. Other caves where human remains were deposited deeper within systems seem to be associated with a continued use of caves as ritual spaces, such as at Grotta di San Michele ai Cappuccini (Skeates 2012, 173–174).

On the Maltese archipelago, Neolithic human remains from unmodified natural caves are rare (Stoddart and Malone 2013, 48). Nevertheless, some Early Neolithic sites, such as Bur Meghez, were used in this way. At this site, up to thirty-nine individuals had apparently been interred as fleshed bodies. There are also a small number of human teeth known from Ghar Dalam cave. At both sites, the human remains were closely associated with animal bone, pottery and stone tools (Zammitt 1930, 58–59). Stoddart and Malone (2013, 48–50) argue that these sites are important as they mark the beginning of a much wider set of practices drawing on the experience of being within caves. Artificial burial caves were constructed at sites such as Xaghra and Hal Safleni, and these sites of the dead found their counterparts in the cave-like properties of the contemporary temple sites of the living, for example at Ggantija and Tarxien. This is an intriguing argument which can be extended to the Late Neolithic of southern Italy, for example at Manfredi (Skeates 2013, 37), suggesting that Neolithic monumentality developed from the experience of using caves for burial.

In Italy and the surrounding islands, we therefore have the earliest evidence for human remains from caves being an important part of the local Neolithic. There appear to have been at least two different traditions. In the early part of the Neolithic, cave burial was similar to burial activity elsewhere in the landscape. Although it usually involved single individuals, it is clear that these burials had complex biographies, and they may not necessarily have been simple primary burials. It is possible that this style of cave burial was directly connected with the adoption of the Neolithic in the region. Later in the period, we can see evidence for the deposition of some human remains in cult caves and also for a practice of collective burial. As was argued for the Balkans, it may be the case that this different style of cave use was connected with an increase in pastoralism and greater settlement mobility in the later Neolithic. It is also possible that these developments acted as precursors for the development of monumental structures used for collective burial in the Mediterranean region.

Southern France and the Iberian Peninsula

Most of the human remains which are known from the earliest Neolithic in southern France come from a small number of caves and rock shelters (Guilaine and Manen 2007, 27). In her review of burial throughout the wider region, Zemour (2011, Figure 1b and see Appendix 2), lists 22 cave or rock-shelter sites in southern France and Corsica with Early Neolithic human remains. There are nine open-air sites from the same period and region with human remains. This parallels the situation in Italy and in the Iberian Peninsula, where cave burials form part of a broader continuum of burial practices. The earliest Neolithic presence in southern France occurs in Provence. The very earliest sites, between 5800 and 5600 BC, contain pottery which is stylistically identical to Italian 'Imprezza' impressed wares (Binder and Maggi 2001, 413–415; Guilaine 2015, 92–95). These sites seem to share other connections with the Italian Neolithic – for example, where houses have been discovered, they were small. The faunal evidence has been used to suggest that sheep and goats were the main animals kept for meat, and there is evidence for the cultivation of both wheat and barley (Guilaine and Manen 2007, 33–37). Across Southern France and the Iberian Peninsula, this earliest 'Italic' Neolithic was succeeded by a full Early Neolithic associated with Cardial Ware pottery (Guilaine and Manen 2007, 37–45). It has been suggested

(Guilaine and Manen 2007, 40) that both hunting and pastoralism also made a relatively large contribution to people's diets in this phase.

Taken at face value, the French data would suggest a greater emphasis on cave burial in the Earliest Neolithic than is the case in the other regions. Typical examples come from L'Abri Pendimoun, Alpes Maritimes, where a number of inhumation burials were discovered. The most recent excavation located two female burials dating to the very earliest pre-Cardial phases of the Neolithic. Both were found in shallow oval graves covered by angular limestone blocks (Binder et al. 1993, 231–143 and see Figure 2.4). Recent publication of a large assemblage of human bone from a later date, at the end of the sixth millennium BC, from a fissure at Mougins-Les Bréguières (Alpes-Maritimes) has shown that some collective burial was also taking place before the end of the Early Neolithic (Provost et al. 2017, figs 5 and 6). Neolithic and Copper Age cave burials from the Iberian Peninsula have been recently reviewed and discussed by Weiss-Krejci (2012 and see Appendix 2). There are six caves in this area with Early Neolithic dates. As with the French and Italian examples, Iberian cave burials at this date seem to be part of a wider tradition

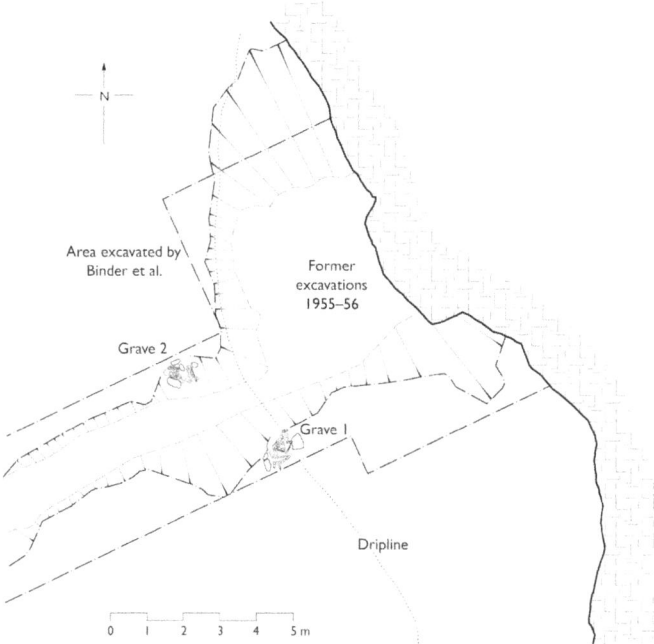

2.4 Earliest Neolithic single graves from L'Abri Pendimoun, Alpes Maritimes (after Binder et al. 1993, Figures 3, 33 and 37).

of burial in simple pits which also occur at contemporary open-air sites (Weiss-Krejci 2012, 119). Despite a previous tendency to see most Early Neolithic burials in the region as single inhumations, Weiss-Krejci (2012, 127) points to substantial collective deposits at Cueva de los Murciélagos, Granada, with the apparent successive inhumation of 'mummified' bodies as evidence of a more diverse range of practices. Successive inhumation in the Early Neolithic also occurred at Gruta do Caldeirão, Alto Ribatejo, where the scattered remains of six individuals were found associated with shell beads and Cardial Impressed Ware (*ibid*).

In southern France, the following Late Cardial and Epicardial phases have been dated to between 5200 and 4800 BC. Beeching (forthcoming) has reviewed this evidence and, although the Epicardial covered a wide geographical range from Spain to the Rhône valley, some overall trends can be seen. In particular, there was an apparent shift away from sheep and goat pastoralism and an increased reliance on a combination of hunting, cultivated cereals and pig-keeping. Settlements also remained small and relatively ephemeral. Further North, in the Jura mountains, Perrin (2003, 737–738) has suggested that hunter-gatherer groups and farming groups were occupying the same areas until as late as 4900 BC. A similar pattern of co-existence has been suggested for parts of the Atlantic coast of Portugal. Zilhão (2001) has pointed to overlapping dates from Mesolithic and Neolithic sites in Estramadura to suggest that colonising groups moved into empty areas of these landscapes.

The Middle Neolithic is a period with extremely ephemeral archaeological traces in southern France, with the exception of some apparent influences from Italy in Provence. Populations in this area presumably remained mobile and continued to hunt, farm and herd animals. Cereal cultivation is attested at some high altitude sites, and faunal remains from both open-air and cave sites show a broad range of domestic species (Bogaard and Halstead 2015, 395–398). A similar pattern can be suggested for northern Spain, where there seems to have been a limited adoption of cereals and domesticated animals alongside hunting during the fifth millennium BC (Zilhão 2000, 147). Detailed analysis of bone stable isotopes from the non-megalithic collective burial at Alto de Reinoso, Burgos, in the northern Meseta, has shown that by the early fourth millennium BC, both cereals and sheep and goats were making significant contributions to people's diets (Alt et al. 2016, 22–23). The major change associated with the latter part of the fifth millennium BC throughout the region was the introduction of megalithic monuments after about 4300 BC (Rojo-Guerra and Garrido-Pena 2012, 22–23). The Middle Neolithic Chasséen culture of southern France, between 4300 and 3500 BC (Beeching et al. 2000, 61), also saw the introduction of more complex funerary monuments and of large open-air settlement sites. Bréhard and colleagues (2010) have analysed the kinds of pastoralism practiced in these areas and conclude that there was a complex seasonal round which incorporated both the large river-terrace settlements and cave sites. Different sites were used at different times of the year in a specialised pastoral system (Bréhard et al. 2010, 186–187; Delhon et al. 2009, 62–63).

There are eight caves in the Iberian Peninsula with Middle Neolithic dates for the deposition of human bone. Therefore, caves seem to have continued to function as mortuary spaces in parallel to the development of megalithic architecture in the same period (Weiss-Krejci 2012, 120). During the Middle Neolithic in France, there is also evidence for a variety of burial practices. The human remains from Fontbrégoua Cave, Var, were discovered in three pits within a large cave which also has evidence for similar pits containing animal bone (Le Bras-Goude et al. 2010, 168–9). Earlier interpretations of the Fontbrégoua human remains focussed on the presence of cut marks and the similarities between the treatment and deposition of human and animal remains to interpret this as evidence of Neolithic cannibalism (Villa et al. 1986). Le Bras-Goude and colleagues (2010, 173–174) used more recent radiocarbon results to demonstrate that one of these pits was used for at least two separate partial interments in the Early Neolithic Cardial phase. Remains from the other two pits dated from the succeeding pre-Chassy phase of the Middle Neolithic. Detailed analysis of the post-mortem treatment of this bone (Villa et al. 1986, 148–154) shows that manual disarticulation, de-fleshing and probably some consumption of the body took place as part of the burial rite. There is also evidence for successive inhumation at this date, for example the twenty-three individuals discovered in crouched postures on the surface of cave deposits at Les Grottes des Barbilloux, Lot-et-Garonne. In some cases, this may be linked to the development of burial in artificial rock-cut tombs. At the rock-shelter site of L'Abri du Pas-Estret, Dordogne, a rock-cut pit contained the successively deposited remains of nine individuals (Beyneix 2012, 225–226).

Individual burials continued to occur in caves in the Late Neolithic, for example at Resplandy Cave, Hérault. There are also cremations at sites such as La Baume des Maures, Var (Vander Linden 2006, 321). Collective burials include Trou de Viviès, Aude, and Can-Pey cave, Pyrénées-Orientales, which incorporates the remains of at least sixty-four individuals (Baills and Chaddaoui 1996, 367). These collective deposits seem to have been the result of a number of different funerary practices. At Aven de la Boucle, there is evidence of successive inhumation of twenty-six individuals in a doline with some rearrangement of earlier burials as part of the process (Vander Linden 2006, 321). At Can-Pey, a combination of osteological and archaeological study suggests the possibility of a secondary burial rite (see Figure 2.5). Bodies may have been placed at the mouth of chamber II, which acted as the first place of burial. There is also evidence of fires having been lit at this time. Once the bodies had decayed, the bones were moved to a secondary deposit in chamber I, around 10 metres away, before finally being moved for the last time into the deepest chamber of the cave (Baills and Chaddaoui 1996, 369–370).

During the Iberian Late Neolithic, there appears to be a substantial increase in the use of natural caves for the deposition of human remains. This is part of a broader trend in this period for the use of megalithic structures, pits, silos and constructed subterranean spaces for burial. There are at least twenty-two sites with well-contextualised radiocarbon dates from this period (Weiss-Krejci 2012, 121–122). The use of natural caves seems to have been particularly common in the fourth millennium, which parallels the suggested

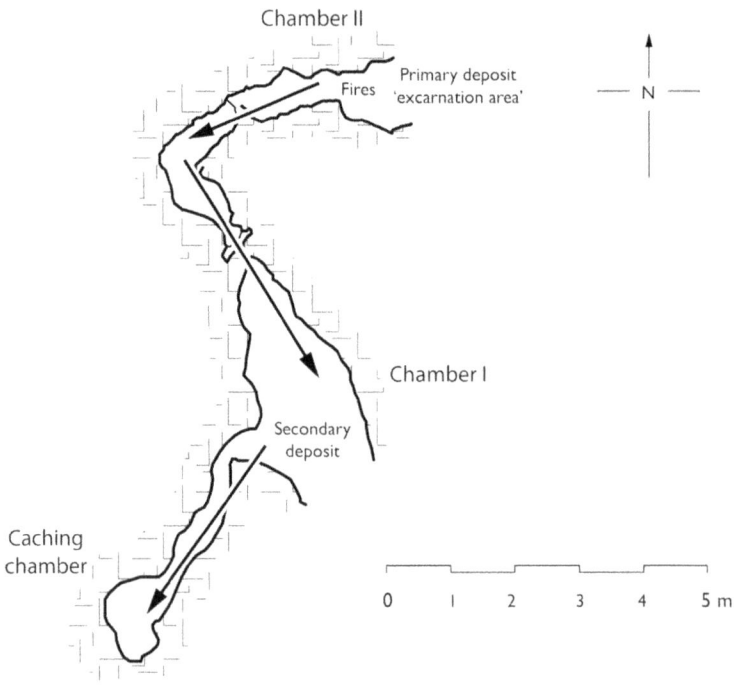

2.5 One of the postulated sequences for multi-stage burial at the Late Neolithic cave site of Can-Pey, Pyrénées-Orientales. After Baills and Chaddaoui (1996, Figure 4).

development (Stoddart and Malone 2013, 48–50) of artificial burial caves, or hypogea, from the use of natural caves further east in Malta and southern Italy. A detailed study of two Late Neolithic rock shelters and one doline in the Ebro Valley appears to show that these sites were exclusively used for the successive inhumation of entire bodies (Fernández-Crespo and de-la-Rúa 2016, 287). Demographic comparison of the individuals from these sites with megalithic graves of the same date and in the same region suggests that more women and children were buried in the karst sites. Fernández-Crespo and de-la-Rúa (2016, 291–295) consider that this was likely to be the result of a deliberate cultural choice within one group who were using both kinds of burial space. Unsurprisingly, given the increased amount of cave burials from this period, there is also evidence for an increased range of burial practices in the Late Neolithic. For example, Bolóres rock shelter, Estremadura, seems to have been used for the earlier phases of a secondary burial rite (Weiss-Krejci 2012, 128–129). Successive inhumation also continued at many sites, and there was evidence for the collection and manipulation of disarticulated bone as parts of the latter phases of secondary burial rites. Well-preserved cut-marked bone from four individuals from El Pirulejo, Andalusia, appeared to have been exhumed from a burial site elsewhere before being re-buried at this cave (Weiss-Krejci 2012, 129).

There appears to have been increased regionalisation and territorialisation after 3500 BC during the Final Neolithic in southern France (Beeching 2002; D'Anna 1995; Gutherz and Jallot 1995). The number of known sites from this period is larger than in the Middle Neolithic, but the settlements are generally small in size. Where buildings have been discovered, they are ephemeral and often only known through traces of wattle and daub. Across the region, there is evidence for complex pastoralism, based on dairying of sheep and goats, the introduction of hulled varieties of cereals, gathering of acorns and fruit and even bee-keeping. Beeching (forthcoming) suggests that throughout most of the Neolithic in southern France, there was only a gradual transition to a full farming economy, which was not completely established until as late as the end of the Final Neolithic. Collective burial (Cummings et al. 2015, 816) also became much more widespread during the Final Neolithic in this region. In the first instance, this took place in natural caves and artificial subterranean hypogea and, after about 3200 BC, increasingly in constructed megalithic gallery graves.

In central France, caves and rock shelters seem to have been used for burial rites associated with the Final Neolithic Artenac culture; in some cases, the human remains were contained within rectangular dry-stone cists (Roussot-Larroque 1984, 160). Further south, six Final

Neolithic individuals were discovered under a limestone cairn within the rock shelter of L'Abri du Moulin du Roc, Dordogne, apparently in a collective primary grave (Beyneix 2012, 231). In this region of France, collective burials in caves are also known from the Beaker period (Vander Linden 2006, 324) and, as was the case further east in the Mediterranean, the Late Neolithic and Early Copper Age also saw the development of a tradition of hypogea. There is also intriguing evidence for curated human bone from four Copper Age caves in Portugal (Weiss-Krejci 2012, 130): Casa da Moura, Gruta do Cadaval, Gruta dos Ossos and Covão d'Almeida. This may be the result of extremely extended multi-stage burial rites but, in view of the evidence for the curation and mummification of human bone elsewhere in the European Bronze Age (Booth et al. 2015), the curation of either dry bone or mummified individuals should also be considered.

If we consider the cave burial evidence from southern France and the Iberian Peninsula as a whole, we can clearly see continuity with the evidence from the Italian Neolithic. Here there were also apparently two different traditions. In the Early Neolithic, burials were often contained within cists or pits and, despite being largely complete, there is evidence of extended burial rites in the creation of many of these deposits (Zemour 2011, 261). This tradition seems to be one of the traits which were introduced into the region as part of the Earliest Neolithic. However, as was also the case further east, the later traditions of collective burial seem to have connections with providing fixed or memorialised points in the landscape for complex pastoralists. These collective burials traditions may, as Stoddart and Malone (2013) suggest for Malta, be some of the earliest burials of this type and therefore mark the beginning of the more widespread use of hypogea and burial monuments later in the region.

Central Europe

A few examples of collective cave burial can be seen in the Neolithic of Central and Northern Europe. Orschiedt (2012, 217–219) has reviewed the evidence from nine sites in Germany, which are concentrated in the southern part of the country and which are largely Late or Final Neolithic in date (see Appendix 2). The Neolithic in this region, and the few cave burials associated with it, may provide a connection between the Mediterranean traditions which I have already reviewed and the more common late fifth and early fourth millennium cave burials in Belgium, Britain and Ireland. The initial development and rapid spread of the LBK Neolithic began in the Transdanubian regions of Central Europe around the middle of the sixth millennium BC.

It spread rapidly north and west, becoming the earliest Neolithic in southern Germany, northern France and parts of the Low Countries. The LBK was characterised by an extremely homogenous repertoire of material culture and settlement styles (Gronenborn 1999, 130–132). As Gronenborn (2007, 79) has noted, the LBK landscape would have been dominated by dispersed and yet highly structured and closely connected arable villages. These settlements seem to have acted as a place where pastoralists, hunter-gatherers and farmers were able to meet. Gronenborn (2007, 79–82) suggests that the LBK village was the place where an 'LBK ideology' around the ritual importance of fertility was communicated and that this communication explains the rapid expansion and extreme homogeneity of the Early LBK Neolithic. In the Paris Basin, Hachem (2000, 310) has shown that, particularly in the Early LBK, the hunting of wild game still contributed an important part of some people's diets.

There is evidence for a range of different practices with human bone even within the small number of sites in Germany. At Jungfernhöhle, Bavaria, there are a large number of fragmented individuals associated with artefacts which range in date from the LBK to the medieval period. Radiocarbon dates on human bone show that there was substantial burial during the Early Neolithic, with some re-use around 3500 BC in the Late Neolithic (Orschiedt 2012, 218). Recent osteological analysis of the Neolithic material by Orschiedt (2012, 217) has shown that the cave was being used for secondary burial. The assemblage was dominated by skull fragments and the major long bones, with the absence of hand and foot bones, vertebrae and more fragile elements of the skull, clearly showing that the bodies had been exposed and skeletonised in a different location.

During the fifth millennium in western central Europe, the uniformity of the LBK was replaced by first the Rössen and then the Bischheim cultural groups (Kreuz et al. 2014, 73–74). Stable isotope analysis of a number of Middle Neolithic burials from southern Germany has shown that diet in the region continued to follow the pattern established in the LBK Neolithic of a relatively flexible use of a broad spectrum of resources (Morseburg et al. 2015, 219). By the Bischheim period, longhouse settlement had been completely abandoned. Traces of burnt daub demonstrate the presence of buildings, but they were of a type which left no sub-surface traces.

Osteological and archaeological evidence from Höhlenstein-Stadel, Baden-Württemberg shows that cave was being used for secondary burial at the beginning of the Late Neolithic (Orschiedt 2012, 218). Interestingly, Late and Final Neolithic human bone from Vogelherd cave (Conard et al. 2004, 200), which has also been analysed by

Orschiedt, shows that at that site there was the successive interment of at least six individuals on the surface at the cave entrance. Carnivore gnawing and the continued articulation of the torso of one burial shows that these bodies were not moved from this cave after deposition. There is also some evidence for the re-use of sites, such as the Blätterhöhle, Westphalia, which had been used for burial earlier in the Mesolithic. Intriguingly, the aDNA and bone chemistry evidence from fourth millennium BC burials at this site shows that two separate populations were using the cave for burial. One group was genetically similar to the earlier Mesolithic burials and appeared to have a diet based on wild resources, especially freshwater fish. The other group had evidence of domesticated food consumption and some genetic evidence for migrant origins (Hofmann 2015, 464). By the Late Neolithic there is also some evidence for single primary burials from caves in southern Germany. For example, a child buried in the upper layers of the Palaeolithic and Mesolithic rock-shelter site of Felsstalle, Baden-Württemberg (Kind 1987, 293–243), has been radiocarbon-dated to the Late Neolithic (Orschiedt 2012, 217).

Belgium

Neolithic cave burial was extremely common in the limestone regions along the Meuse valley. A recent estimate (Crombé and Robinson 2014, 564) suggests that there are over 220 caves with human remains in this region, the vast majority of which probably date to the Seine-Oise-Marne Late Neolithic period. The archaeology of this group of caves has been reviewed by Cauwe (2004), who has established that cave burial in this area was practiced in the Michelsberg Middle Neolithic in addition to the Late Neolithic. He suggests that these burials form a coherent set of rites covering both periods from the early fourth millennium to the middle of the third millennium BC (Cauwe 2004, 220–221). Caves seem to have been used for the collective burial of relatively small groups of people. Most of the sites reviewed by Cauwe (2004, 220) have a minimum number of between fifteen and twenty individuals in each cave. The most common items of material culture associated with the burials are worked stone and some worked animal bone, with pottery being extremely rare. Burial practice was variable, but all caves seem to have evidence for repeated successive inhumation (Cauwe 2004, 219–220).

Around 4300 BC, the Michelsberg culture developed in Belgium and northern France, and from there it spread into southern Germany. Detailed stable isotope analysis on human bone has provided some useful background on the daily lives of people in this area. During this period, there was a relatively high consumption of wild foods, especially freshwater fish (Bocherens et al. 2007, 19). There is also archaeobotanical evidence for a reduction in the range of cereal crops being grown, which may be linked to an increased reliance on stock breeding (Kreuz et al. 2014, 93–95). Settlement evidence for the Michelsberg Neolithic is dominated by pits and a very few small, sunken-floored buildings. After the Michelsberg period, from around 3300 BC, there is evidence for a number of different Late Neolithic groups. The Seine-Oise-Marne Late Neolithic of Belgium and Northern France is particularly important for this study, as most of the documented cave burials in the Meuse Basin belong to this group. The study by Bocherens and colleagues cited here (2007, 19) shows that, in the Meuse basin at least, Late Neolithic people in Belgium were eating both cereals and domestic mammals.

In the Middle Neolithic, there were relatively fewer cave burials, and these early sites are often the ones with the lowest number of individuals. This may suggest that individual burial was an early rite. The Trou de la Heid, Liège, contained the extremely fragmentary remains of one adult and one child, dated to 3380–3530 BC. The individual burial at Chauveau CH1 is dated even earlier, to 3900–3650 BC (Toussaint and Becker 1994, 78–82). The Abri des Autours, Namur, is one of the few examples where a cave in this region can be demonstrated to have been used for the final phases of a secondary burial rite. A deposit of the fragmentary remains of three adults and six juveniles was discovered near the entrance of a cave which had previously been used for burial in the Early Mesolithic (Figure 2.6). This collective deposit has a radiocarbon date which would calibrate to between 4320 and 3980 BC. Two of these burials seem to have been deposited immediately after death, but the other seven were either extensively rearranged at some time after burial or were moved into the cave from an intermediary period burial site elsewhere (Polet and Cauwe 2007, 74–84).

There are examples of cremated bone being discovered amongst the majority of the unburnt bone. This occurred both during the Middle Neolithic, at sites such as Trou du Frontal, Furfooz; and during the Late Neolithic, for example at Trou des Blaireuax. There is also evidence for cut marks made by stone tools on human bone from sites such as Caverne B, Hastière, where it would date to the Middle Neolithic; and Fisure Jacques, Chanxhe, which has a Late Neolithic date. This has been sometimes interpreted as evidence of cannibalism, but in this case, it is more probably evidence for an extended funerary rite involving de-fleshing (Cauwe 2004, 220). Although burial normally took place in unmodified natural caves, there are some examples where low dry-stone walls or pavements were used to separate burials from the rest of the caves. As well as the Middle Neolithic example of Abri des Autours (see Figure 2.6), there is a Late Neolithic example from Trou des Blaireaux, Vaucelles. In some cases, it has been suggested that caves were closed at the end of their funerary use. One good example is the Trou du Frontal, where a large slab had been dragged vertically in front of the cave. In the Late Neolithic, Grotte Triangulaire, Ramioul was sealed with a dry-stone wall (Cauwe 2004, 219–220).

Cave burials in the Belgian Middle and Late Neolithic seem, therefore, to have been predominantly collective burials. There is evidence for a range of different practices, but it is clear that even the earliest burials are much too late to be directly connected with the introduction of the Neolithic into this region. They are much more likely to

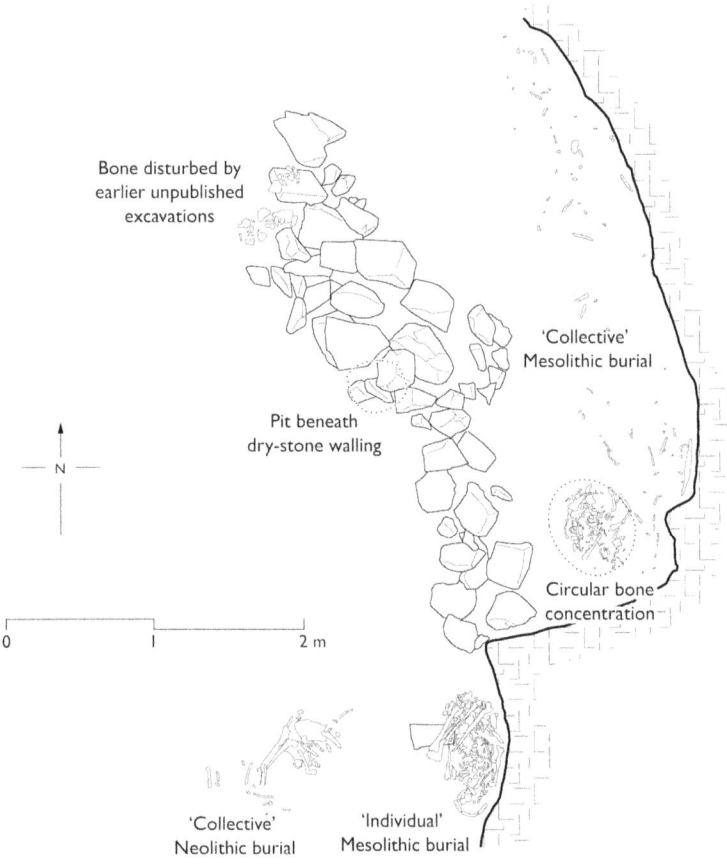

2.6 Cave burials of various dates in the Abri des Autours, Namur (after Polet and Cauwe 2007, Figure 2).

represent a local variant of the practice of collective burial in natural places, hypogea and monuments which I have already noted further south in Europe. The few Middle Neolithic Michelsberg burials may be another example of the link noted in other regions between the adoption of cave burial and a pastoral or semi-mobile lifestyle. However, the evidence for this is not particularly strong, and most of the cave burials in this region would fall into the later Seine-Oise-Marne period. In this case, the evidence suggests that the people carrying out the burial would have been settled farmers.

Cave burial in Ireland

Neolithic cave burial in Britain is discussed in detail in Chapters 5, 6 and 7 of this book. The archaeology of Neolithic cave use in Ireland has been extensively reviewed by Marion Dowd (2008 and 2015). Human bone which has been radiocarbon-dated to the fourth and third millennia has been found in eighteen Irish caves (see Appendix 2). With the dates currently available, there seems to be a peak of activity in the middle of the fourth millennium (Dowd 2015, 95), around 150 years after the beginning of the Irish Neolithic (Whitehouse et al. 2013, 185–188). Although some of these sites are reused in the Bronze Age, there is little evidence of Late Neolithic activity (Dowd 2015, Figure 5.2).

Neolithic practices spread into Britain and Ireland early in the fourth millennium BC. There is evidence at this date for both arable agriculture and for relatively substantial buildings. The date and nature of the Irish Neolithic has been recently reviewed by Whitehouse and colleagues (2013). They would see a rapid beginning to the Irish Early Neolithic around 3720 BC. Large numbers of relatively small rectangular houses were constructed over the following 100 years, sometimes clustered together in groups of five or six. The archaeobotanical evidence seems to show that the people living in these houses were carrying out intensive cereal agriculture in small, intensively tended 'garden' plots (Whitehouse et al. 2013, 196–199). Irish causewayed enclosures were also built during this period (Whittle et al. 2011, 383), which suggests that there was a desire for people to come together seasonally in larger groups than those who lived in the excavated settlements. This was also the date at which the earliest megalithic tombs in Ireland began to be used (Whitehouse et al. 2013, Table 3), showing that people's connections to the landscape were also drawing on the visible and permanent presence of the dead.

There is evidence for a range of different cave burial rites in Ireland, sometimes in the same sites. For example, at Annagh Cave,

Limerick, five individuals have been identified (Figure 2.7). Two of these, Annagh 1 and 2, seem to have come into the cave as fleshed bodies and were deposited as crouched burials close to the cave wall. Nearby was Annagh 3, which, although it superficially resembled another crouched burial, was made up of the rearranged major bones of a disarticulated skeleton, indicating a use of the cave for the later phases of multi-stage burial rites. Individuals 4 and 5 from Annagh show a different phase of secondary burial. The small bones and extremities of these two individuals survived in a deposit to the north-east of the other burials, presumably after the larger skeletal elements had been removed for burial or curation elsewhere (Dowd 2015, 98–100).

The 'house horizon' in Ireland was followed, after about 3600 BC, by a Middle Neolithic in which settlement was more widely dispersed

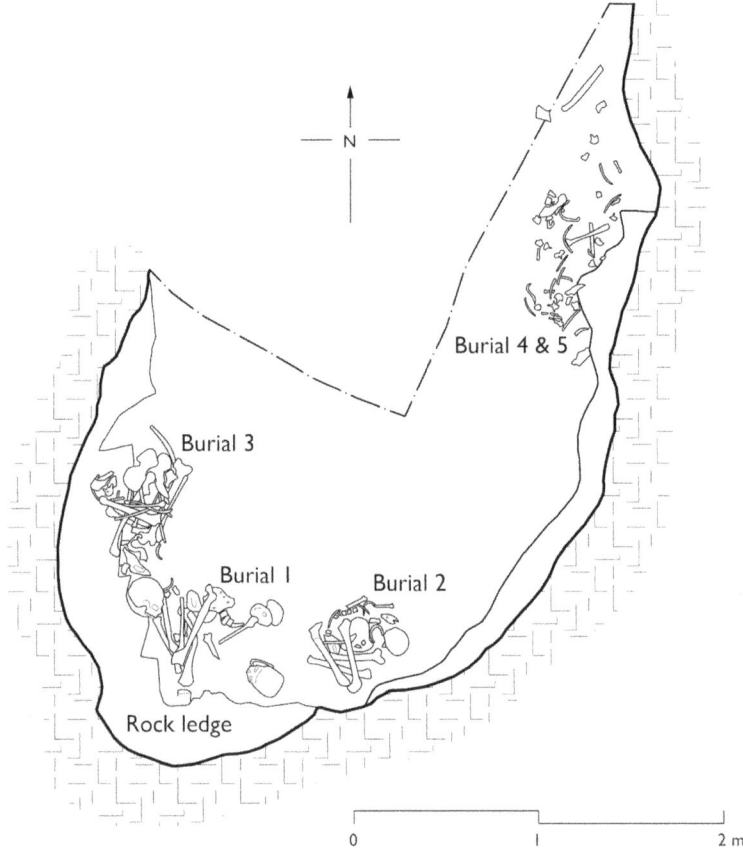

2.7 Neolithic human remains from Annagh Cave, Limerick (after Dowd 2015, Figure 5.3).

and left more ephemeral traces. At the start of this period, there seem to have been changes in the climate and environment, with the onset of generally wetter conditions. Possibly as a response to this change, there was a decline in the amount of the landscape cleared for farming and a suggestion that gathered wild plants became an important part of people's diets once more (Whitehouse et al. 2013, 199–200). It has been suggested that a similar shift towards more ephemeral settlement and away from arable agriculture occurred in Middle and Late Neolithic northern and western Britain. This was certainly true of Wales after 3000 BC (Peterson 2004), and Stevens and Fuller (2012, 712–714) suggest that cereal agriculture in all of Britain declined markedly at around 3350 BC and that Middle and Late Neolithic people were almost exclusively pastoralists.

Cave burial in Ireland seems, therefore, to be primarily an Early Neolithic phenomenon. However, the dating evidence cited here suggests that it was not directly connected with the transition to the Neolithic. The situation in Britain seems to be different, and the relationship of cave burial practices to the earliest Neolithic is considered in more detail in Chapter 5. In both countries, there is evidence that monuments functioned as fixed points associated with the dead within the landscape (Cummings 2017, 130; Whitehouse et al. 2013, Table 3; Whittle et al. 2011, 383), and it is possible that cave burial was fulfilling a similar role at this date. At present, the Irish data shows very little evidence for later Neolithic cave burial at all. This is another area where there is some difference apparent with the situation in Britain. The Middle and Late Neolithic evidence for cave burial rites will be discussed in more detail in Chapter 7.

A connected continent

From this review of the archaeology of Neolithic human remains from caves across Europe, several important broader themes are apparent. Where cave burial was associated with the first adoption of a Neolithic way of life, then it is clear that this was primarily a Mediterranean phenomenon (see Figure 2.8). There is almost no evidence for human remains from caves from the LBK Early Neolithic. Where there were large numbers of burial caves from these regions, as for example in the Meuse Basin in Belgium, they date to the Middle and Late Neolithic. Interestingly, many of these areas do have large numbers of Mesolithic human remains from caves (Bocherens et al. 2007, 11; Orschiedt 2012), so that we can argue that in the northern parts of Europe, one of the hallmarks of the adoption of a Neolithic way of life is the abandonment of cave burial practices. However, in southern

Europe, we do have evidence from most of Italy and from southern France of a significant number of caves with human remains from the very earliest phases of the Neolithic (see Figure 2.8). Drawing on the work of Zemour (2011) and Robb (2007), discussed in more detail here, it is clear that what was important was that caves were used for a highly variable set of funerary practices, all of which also took place at other, non-cave, locations. Therefore, in the Early Neolithic of Italy and France, there was no 'cave burial' practice as such. Rather, caves were one of a range of available and significant locations which could be drawn upon for a number of different funerary rites.

On the other hand, we can see evidence for cave burial as a practice which is more strongly associated with the developed phases of Neolithic activity in most of the limestone regions of Europe. Where cave burials occurred in significant numbers at the start of the local Neolithic sequence, as they do, for example, in southern France, then they also occurred in even larger numbers later in the period. The obvious exceptions to this statement are the cave burials in Britain and Ireland, where the majority of burials took place early in the local Neolithic sequence. Examining the Europe-wide evidence for the fourth millennium BC (see Figure 2.9), then we can see strong indications that cave burial was particularly common in this period.

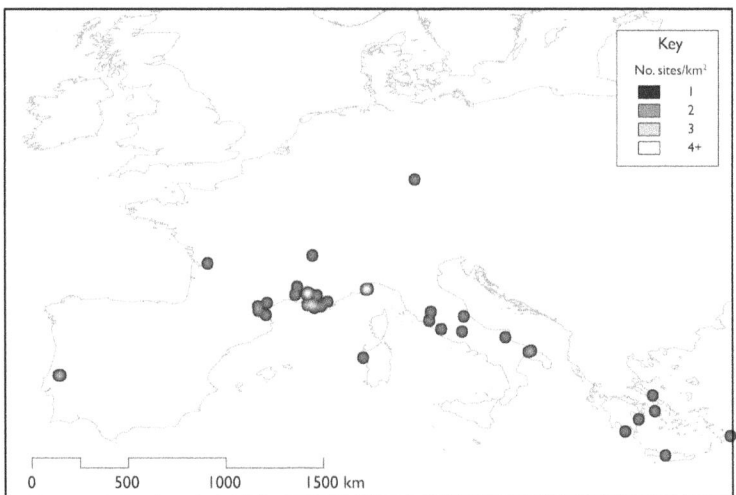

2.8 Caves where human remains were deposited between 5500 and 5000 BC, showing the significant number of such sites from the French and Italian Early Neolithic but also the marked absence of such sites from the Eastern Adriatic coast. The data are presented here as a heatmap. See Appendix 2 for the original sources for this data (total number of sites = 42). The base mapping includes data licenced from © EuroGeographics.

In praise of limestone 37

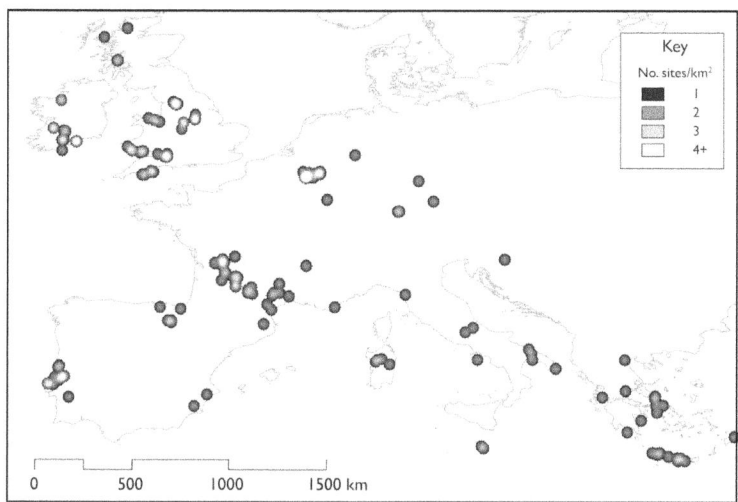

2.9 Cave burials in Europe dating to the fourth millennium BC. The data are presented here as a heatmap. See Appendix 2 for the original sources for this data (total number of sites = 178). The base mapping includes data licenced from © EuroGeographics.

Therefore, the British and Irish evidence could be seen as reflecting this general trend, rather than necessarily being directly associated with the local transition to the Neolithic. If we wanted to identify a time when the use of caves for burial was at its height across Europe, then the beginning of the fourth millennium BC was that time.

This increase in cave burial in the fourth millennium BC is probably also relevant to the origin of megalithic burial sites in the Mediterranean. It has been suggested that there was a development from collective burial in caves leading to collective burial in hypogea and, finally, to the construction of megalithic monuments. As discussed here, this hypothesis has been applied to the origins of monumentality in Malta and Italy (Stoddart and Malone 2013), the Iberian Peninsula (Oosterbeek 1997, 70–71) and southern France (Beyneix 2012, 224). Individual regional examples of this process appear convincing but, as Weiss-Krejci (2012, 121–122) has pointed out, in the specific cases of Spain and Portugal, there were probably more complex relationships between all three classes of site. The detailed and modelled radiocarbon evidence is not yet available to allow us to state definitively that all megalithic burial sites owe their origin to an earlier practice of cave burial. It is more likely that, from the middle of the fifth millennium BC onwards, there was an increase in collective burial in a variety of spaces. In each region of Europe, a historically contingent version of this trend led to increases in monument building, cave burial or the construction of hypogea. A good example of this process is provided by Scarre's (2002) discussion of the adoption of monuments in the mid–late fifth millennium BC in north-western France.

The review of European burial practice presented here has also shown the diversity of practices present. This is especially the case early in the regional sequences in the western Mediterranean, Belgium and Britain and Ireland, where a wide range of practices can be identified from Early Neolithic caves. These include single and double burials, secondary burial rites, successive inhumation, curation and circulation of body parts and possibly mummification. Figure 2.10 shows the total number of caves in use for burial in each 500-year period of the European Neolithic. The examples reviewed here suggest that the increase in the absolute number of burial caves in use from the mid-fourth millennium onwards coincides with an increasing focus on collective interment at the expense of other rites.

There may have been connections between the kind of Neolithic present in different regions at different times and the relative popularity of cave burial. The *habitus*, to adopt Bourdieu's (1977, 73–95) terminology, involved in different kinds of daily farming activities would have structured people's understandings of the world, of the passage of time and of their relationships with each other. We may be able to see a broad-scale reflection of this when we compare their burial choices with what we know of the details of their everyday lives. In the Eastern Adriatic, it is only following the shift to a complex pastoral economy after around 4800 BC that we see significant numbers of cave burials (Mlekuž 2005, 42–43 and see Figure 2.11). A similar case can be made for the adoption of pastoralism in Macedonia after

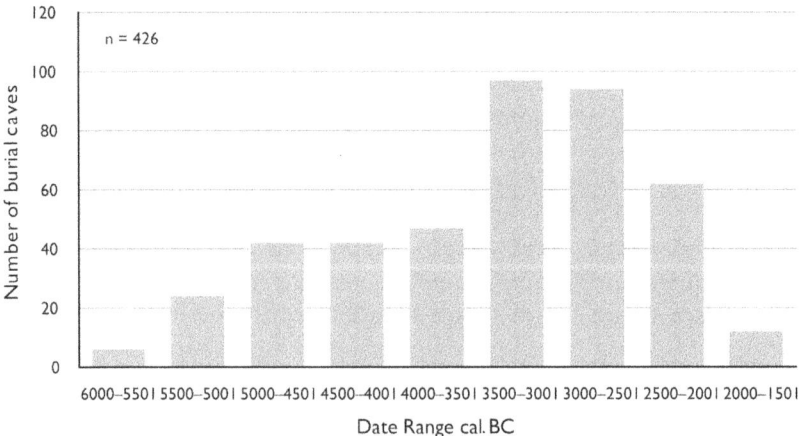

2.10 Caves with human remains for each 500-year interval during the European Neolithic. See Appendix 2 for the original sources for this data.

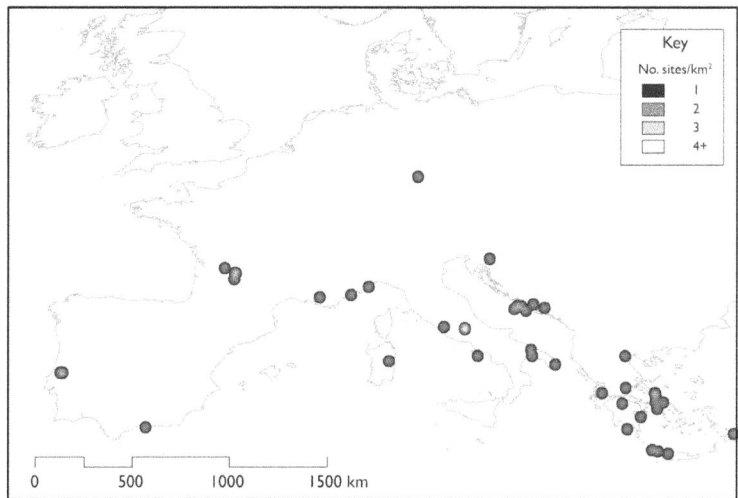

2.11 Caves with burial activity dating between 5000 and 4500 BC. Note the large number of sites in Greece and the Eastern Adriatic of this date, which coincides with the development of pastoralism in these regions. The data are presented here as a heatmap. See Appendix 2 for the original sources for this data (total number of sites = 42). The base mapping includes data licenced from © EuroGeographics.

around 4900 BC and in southern Greece after 4500 BC (Demoule and Perlès 1993, 398–400 and see Figure 2.11).

A similar relationship may also lie behind the large numbers of cave burials in Southern France in the Mid–Late Neolithic (see Figure 2.9) which also initially coincided with an extremely mobile and ephemeral phase in the local Neolithic (Beeching, forthcoming) and then with the complex pastoral system of the Chasséen Late Neolithic (Bréhard et al. 2010, 186–187). In all these cases, it may be that funerary caves were able to provide a fixed, memorialised point in the landscape for dispersed and mobile pastoralists. The idea that the presence of the dead provided a fixed point in the landscape may have applied in other areas too, in Ireland and Belgium, for instance, but in these cases there is a less directly provable correlation between the adoption of a complex pastoral system and the beginning of cave burial.

All of these interpretations provide us with a background of possibilities, but a more detailed and nuanced view requires a much more densely described and analysed dataset. In the remainder of this book, I shall be attempting to apply some of these insights at a local scale to human remains from British Neolithic caves. However, before discussing this detailed evidence, there are a number of theoretical and methodological details which need clarification. Thus far, I have been describing the kinds of funerary practice which I think took place by adopting, relatively uncritically, the interpretations of the original excavator. Terms like 'collective burial', 'multi-stage burial', 'excarnation' and 'successive inhumation' have been used without any attempt to either define them precisely and consistently or to investigate what kinds of funerary rites and beliefs may have been behind them. Therefore, in the next chapter, I will review the ethnography and osteoarchaeology of multi-stage burial rites, the geoarchaeology of caves and the taphonomy of bodily decomposition to provide a consistent set of interpretive models for cave burial practices.

3

Gestures and positions

In Nicolas Cauwe's (2004, 220) review of the Neolithic burials from the Meuse basin, Belgium, he uses the phrase *'gestes posés sur les cadavres'* to refer to the analysis of the way in which bodies are deposited. While checking my literal translation of this as 'gestures and positions of the bodies', I noticed my dictionary gave several examples of the idiomatic use of *gestes posés* to mean 'the rules of the game'. 'The rules of the game' for the bodies seems to me an excellent summary of the embodied nature of burial practice while at the same time reminding us of the importance of repeated practice in reiterating particular kinds of funerary rite. I will examine the way that funerary rites in caves were created and remembered in more detail in Chapter 4. This will be from the perspective that the human agency of the living, the taphonomic agency of the corpse and the material agency of the cave were united in Neolithic cave burial practice. To understand this process, it is therefore essential that we have a clear set of criteria for describing and interpreting each set of evidence. As discussed in Chapters 1 and 2, human bone from caves and rock shelters has been linked to other Neolithic burial evidence, especially those from within chambered cairns (e.g. Barnatt and Edmonds 2002; Beyneix 2012). Schulting (2007) has also discussed caves within a wider review of 'non-monumental' burial in the period. The primary reason that burials from all of these contexts have been regarded as connected is because they are usually discovered in a more or less fragmented state. Therefore, if we want to interpret cave burials, we need to discuss them in the context of wider debates about Neolithic burial rites. In particular, the processes by which bodies became fragmented and co-mingled need to be analysed and compared.

A history of the interpretation of multi-stage burial practices in the Neolithic

The funeral rites which lay behind collective disarticulated burials have been reconstructed in different ways in the past. Antiquarian accounts, for example John Thurnam's nineteenth-century excavations of Wiltshire long barrows, tended to assume that these sites contained single mass-burial events. Thurnam's interpretation of the six individuals he excavated from the terminal chamber at West Kennet, Wiltshire, was that they were the remains of a chiefly burial surrounded by sacrificed retainers (Thurnam 1860, 414–416). By the middle of the twentieth century, the predominant explanation for such deposits was that they were the results of successive burials in a communal grave or ossuary. For example, the nine individuals recovered from the Lanhill Long Barrow, Wiltshire, were recognised as having been placed successively into the chamber (Keiller et al. 1938, 128–129). The excavators provided a detailed description of the disposition of all of the skeletal elements. This allowed them to reconstruct the funeral rite as the successive crouched inhumation of individuals. The most recent inhumation was discovered as an intact crouched inhumation, with the bones of earlier burials moved to the back of the chamber. There, they had been placed in 'symmetrical' arrangements with some re-articulation of crania and mandibles (Keiller et al. 1938, 125–127). Keiller and colleagues (1938, 128–129) explicitly considered, but rejected, the alternative hypothesis that the Lanhill burials were the final resting place of bodies that had become skeletonised through a secondary burial rite involving other locations. They drew on analogies with Mycenean Tholos tombs to suggest that British chambered tombs were 'family sepulchres' used over a 'considerable period of years'.

By the time of Paul Ashbee's (1966) excavations at Fussell's Lodge Long Barrow, Wiltshire, interpretations had shifted again. At this site, the highly fragmented remains of between fifty-three and fifty-seven individuals were found beneath a flint cairn at the east end of the long barrow. The cairn covered the remains of a timber mortuary structure which had contained the bone. Drawing on the state of the bone and the very partial representation of most of the individuals, together with the evidence for rodent gnawing, Ashbee (1966, 37–42) suggested that the disarticulated remains had been exposed before their burial. The lack of bone outside the main burial area led Ashbee to suggest (1966, 38) that an external site was used as the location of this exposure phase. Once the bones had become de-fleshed, the disarticulated remains were gathered up and placed into the long barrow chamber. Ashbee (1966, 38–42) broadened this interpretation to postulate a secondary funerary rite for most long barrows which linked them to the human bone found at causewayed enclosure sites. In this model, a long-term secondary burial rite was assumed to be the norm for the British Early Neolithic, with distributed pieces of disarticulated bone used at a variety of sites before finally being laid to rest in long barrows and chambered cairns.

This multi-stage model for Neolithic burial rites became increasingly influential in the latter part of the twentieth century, as seen, for example, in Edmonds' (1999, 58–67) view of the circulation of ancestral human remains as part of the experience of daily Neolithic life. This was, in part, because of a wider knowledge of the comparative ethnography of similar secondary funerary rites around the world. In their detailed reconstruction of the burial rites at West Kennet, Thomas and Whittle (1986, 135) drew upon the work of Van Gennep on rites of passage. They used this to interpret a difference between successive inhumation within the chambers, which they saw as having taken place at West Kennet, and more public forms of excarnation associated with long barrows, following Ashbee's reconstruction of the Fussell's Lodge rites. Ethnographic analogy was also used in this report to discuss evidence for the circulation of bone outside the tomb. Drawing on the work of Hertz on Indonesian burial practice, and an example from Strathern's work on the curation of bone (Thomas and Whittle 1986, 148), they argued that, although the bodies were originally skeletonised by successive inhumation, some skeletal elements had subsequently been removed from the tomb. Therefore, both the deposition and circulation of human bone had performed an important symbolic function.

More recent studies, especially of Cotswold-Severn cairns, have returned to interpreting burial rites as the successive interment of many bodies in the same chamber, seeing the wider circulation of bone in secondary burial rites as less plausible. The human remains from the two lateral chambers at Hazelton North, Gloucestershire, were interpreted by Saville (1990, 250–252) as the result of successive interment, largely on the evidence of the presence of an almost completely articulated individual as the last deposit in the north passage. Although some bones were probably removed from the chambers, Saville (1990, 251) did not regard this as convincing evidence for the circulation or symbolic importance of disarticulated human bone. At Wayland's Smithy I, Wiltshire, the fragmented and co-mingled human remains of 14 individuals were discovered within a timber mortuary structure similar to the one discovered by Ashbee at Fussell's Lodge. The original excavator, Richard Atkinson, had interpreted these bones, following Ashbee's model, as the final deposition at the end of a secondary burial rite. However, study of the detail of the mortuary deposits has shown that a sequence of deposition can also be seen here. Drawing on work in forensic anthropology (Haglund and Sorg 1997; Haglund et al. 1988), it was possible to demonstrate that at least the last five individuals were deposited in the mortuary structure as fleshed bodies, and it is likely that this was true of the majority of the burials (Whittle et al. 2007, 104–106 and see Figure 3.1). Two individuals were certainly in an advanced state of decomposition when they were deposited. However, rather than being interpreted as evidence for the circulation and curation of human bone, it was suggested that these were the remains of people killed at scenes of conflict (Whittle et al. 2007, 107). Schulting and Wysocki (2005, 127–128) have proposed that the combination of perimortem trauma, canid scavenging and associated arrowheads suggests that battlefield recovery rather than formal excarnation and bone curation lies behind many examples where decomposed parts of individuals were buried in collective deposits.

The idea that successive inhumation was the normal burial rite in British Neolithic collective deposits has been increasingly influential in studies of chambered cairn mortuary practice. Burials at the Quanterness chambered cairn, Orkney, were interpreted (Renfrew 1979, 166–168) as the product of a secondary burial rite involving the circulation and curation of human bone. This site has subsequently been reinterpreted (Reilly 2003, 149; Schulting, Sheridan et al. 2010, 9), with both re-assessments substituting successive inhumation for the original suggestion of a multi-stage excarnation rite. Nevertheless, it is important to stress that there is often evidence for active choices being made about the re-arrangement of bone even within those chambered cairns where successive inhumation seems to be the main burial rite. For example, Wysocki and Whittle (2000, 595–601) discuss the processes behind the arrangement of human remains in the chambered tombs of the Black Mountains of Wales. At both Penywyrlod and Pipton, there are clusters of disarticulated bone which can be interpreted as an attempt to recreate discrete individuals. Each cluster actually contains the remains of a number of different individuals, but the elements present are approximately the ones required to make up a complete individual. Therefore, in contrast to the interpretation offered by Saville (1990, 251) for Hazleton North, there is evidence that the dried bones of earlier interments continued to be important during the process of successive inhumation.

One of the important distinctions between these contrasting interpretations would have been to do with the temporality of the funeral. There are certain physiological constraints, discussed in more detail

Gestures and positions 45

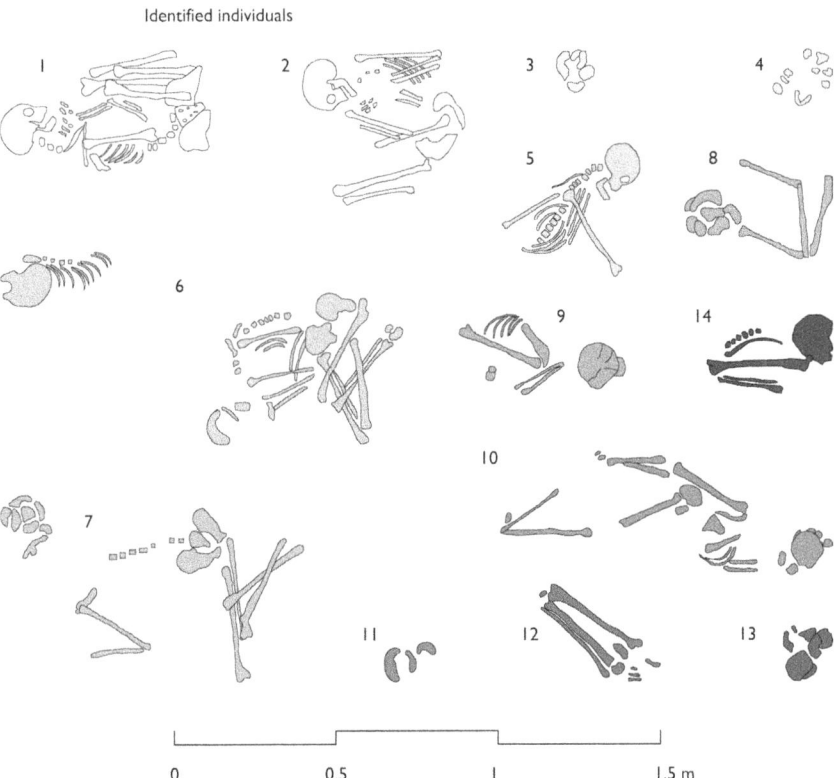

3.1 Michael Wysocki's reconstruction of the burial events in Wayland's Smithy 1 (after Whittle and colleagues 2007, Figure 2).

later in this chapter, which would have influenced how long bodies took to become skeletonised. However, one key factor would have been social decisions about how frequently and for how long a funerary space was used. Many of the interpretations offered by earlier writers assumed that chambered tomb burials took place over an extended period of time. More recent studies of the date and duration of activity at chambered tombs have been based on Bayesian statistical modelling. These studies have shown that for Wayland's Smithy (Whittle et al. 2007, 117–118), burial was both relatively late in the Early Neolithic and probably only lasted for one or two generations. Similar short chronologies have been proposed for Ascott-under-Wychwood, Gloucestershire (Bayliss, Benson et al. 2007), and for West Kennet, Wiltshire (Bayliss, Whittle and Wysocki 2007). Given the range of burial practices which may have taken place at Wayland's Smithy (Whittle et al. 2007, 104–106), it is likely that different post-mortem treatments were taking place simultaneously. Against this evidence, however, it should be noted that the preferred interpretation of the Medway group megalithic tomb at Coldrum, Kent, does suggest that there were episodic burials at that site over several generations after the initial use of the site (Wysocki et al. 2013, 21). It should also be remembered that although burial may have ceased, this does not mean that a monument and the skeletons within it stopped being significant to people.

There were clearly different Neolithic rites at different times and places, all of which could produce a collective, disarticulated burial deposit. There have also been fashions in the interpretation of these deposits, with secondary burial or successive inhumation in favour at different times. It is noticeable that authors who reconstruct burials as successive inhumations have tended to draw more upon the osteological and taphonomic literature to support their arguments, whereas those who advocate a multi-stage rite have made more use of ethnographic analogy. Greater or lesser weight has also been given by various authors to evidence about the circulation and movement of bone. The variability in both the range of evidence and the range of interpretations suggests a need to broaden the discussion. Both taphonomic processes and deliberate actions by the people carrying out the burials were clearly an important part of all of these different rites. The relationship between human intervention, natural processes and time in these extended burials is one which has been discussed from a number of different standpoints in the anthropological literature.

The journey: Ethnographies of multi-stage burial

The ethnography of collective and multi-stage burial is clearly an important part of any attempt to interpret these deposits. These are rites which have been the focus of both archaeological and anthropological research for a number of regions and periods. Two influential studies have been Metcalf and Huntington's (1991) review of the anthropology of transitions around death, and the collection of essays edited by Bloch and Parry (1982) on the power of death as transformation. Both of these studies are founded on the pioneering work of Robert Hertz at the beginning of the twentieth century. Metcalf and Huntington (1991, 33–35) took from Hertz the insight that death, in most cultures of the world, is not conceived of as an instantaneous process. Drawing on Indonesian examples, particularly from Borneo, Hertz developed the concept of the 'intermediary period': the time when the corpse is conceived of as neither fully alive nor finally dead. Hertz (1960) suggested that the decomposition of a body acts as an indicator of the state of the soul. Bodily decomposition shows how the soul travels on a journey between its former existence in a fleshed body within the social world of the living and its ultimate resting place with the ancestors, as it is reduced to dry bones. For Hertz (1960, 201–202), this journey is the central interpretive concept which links together all kinds of extended or multi-stage rites. Cremation, embalming, exposure burial and secondary burial can all be treated as long-term processes aimed at managing and controlling the rate and nature of bodily decomposition and, therefore, the progress of the soul. When the soul has reached the appropriate stage, this point can be marked by a final ceremony, which usually includes the secondary deposition of the human remains in a different location, and which has the social effect of freeing the living mourners from the taboos that they were placed under during the intermediary period. This final ceremony also has the effect of marking the re-birth of the soul of the dead person into a new state of being, the final proper resting place of the dead, from which they were excluded during the intermediary period. Therefore, the final ceremony is a point of release for both the dead and the living (Hertz 1960, 204–206).

Metcalf and Huntington (1991, 43–75) attempted to investigate the extent to which universal patterns could be recognised in human responses to death. The nature of death as a transition from one state to another and the extended process of secondary burials were two areas which they recognised as relevant to the cross-cultural study of human responses to death. Fortunately for the coherence of this book, many of the examples they synthesised involved the use of caves in long-term burial processes. The Toradja of the central Celebes in Indonesia was one of the groups whose ethnographies contributed to Hertz's original account of the intermediary period. Metcalf and Huntington (1991, 99–100) provide a more up to date review of later twentieth-century fieldwork with this group. When a Toradja person dies, their corpse is moved to a hut which has been built away from the village. While it is decaying, it is looked after by a slave, who has to clean up the liquids of decay but also guard the body to keep it from being stolen by witches. At the point of death, the Toradja believe that the soul changes into a dangerous spirit. This spirit can be heard, as low grumbling noises, and smelt, as the scent of decomposition, and it has the power to burn skin and cause diseases. The living relatives of the deceased are also regarded as problematic during the intermediary period. A widow, for example, is confined within a screen of mats and kept on a highly restricted diet. The intermediary period ends with a mass final ceremony held every few years for everyone who has died since the last such ceremony. Shamans summon the unquiet spirits of the dead and instruct them in how to travel to the underworld. As the dead are assisted into the underworld, then their bones are removed from the funerary huts, brought back to the village and collected into a bundle. As the ceremony finishes, the bone bundles are placed into small wooden boxes and then into a cave which already contains older bone bundles of their kin. At the end of the final ceremony, both the problematic nature of the dead and the living mourners have been resolved. The slave who guarded the corpse is freed, although they will be shunned by other people. Caves, in this Toradja example, function as the final resting place for skeletonised and disarticulated, but still identifiable, individuals.

Among the Bara of southern Madagascar, Metcalf and Huntington (1991, 113–130) noted funerary rites which, in many ways, mirrored those of Indonesia. Here too, there was an intermediary period and a final ceremony. However, what was lacking was any explicit narrative about a spiritual journey for the dead person's soul. Here, the physical decomposition of the corpse seemed to be a focus of an intermediary period which addressed structural oppositions within Bara society. Interestingly, although probably coincidently, in this example, caves are used in the intermediary period rather than for the final ceremony. In the three days after death, the mourning and preparation of the body are carried out along strongly marked gender divisions. At the end of this period, a procession, which takes the form of a stylised competition between male and female youths, will take the body to a burial mountain. This may be several miles away from the village. The coffin is placed into a small opening in the burial cave, which is then sealed with rocks. Following this first burial ceremony, there will be a gathering, a pre-planned event that occurs each year at the same season. The final re-burial ceremony in Bara culture can be delayed for a long time, but it must take place before all of the social obligations on the living can be discharged. For example, a widow is not free to re-marry until after re-burial has taken place. The dried bones of the body are removed from the temporary cave and carried in another procession. They are cleaned, re-dressed and re-wrapped before being placed into a communal casket with between ten and fifteen other individuals. These communal caskets are arranged according to kin affinities, with the entire communal tomb acting as a map of genealogical relationships. Despite the apparent focus throughout the process on the social structure of the living, Metcalf and Huntington (1991, 129–130) point to the way in which the 'drying' out of the 'wet' corpse works as a symbol within the wider structural system of Bara society (Figure 3.2). The problem of death, for the Bara, is a problem of imbalance between order and vitality. As the corpse dries, the way is open for the funeral rituals and gatherings to restore the balance within the society.

Bloch and Parry (1982), although they were similarly influenced by the work of Hertz, analysed protracted funerary rites in a different way. In their work, death is intimately connected with fertility. In many societies around the world, according to Bloch and Parry (1982, 7–9), life is a 'limited good'. There is only so much 'life force' to go around, and therefore, in order that new things and people should be born, death is a necessity. This regenerative view of a cycle of birth and death depends on a cyclical conception of the passage of time. Bloch and Parry (1982, 10–11) draw on the work of Edmund Leach to suggest that this view creates a fundamental tension between the contingent flow of events, such as an individual death, and the more ideological concepts of enduring cyclical social order. They suggest that multi-stage burials should be thought of as attempts to reconcile

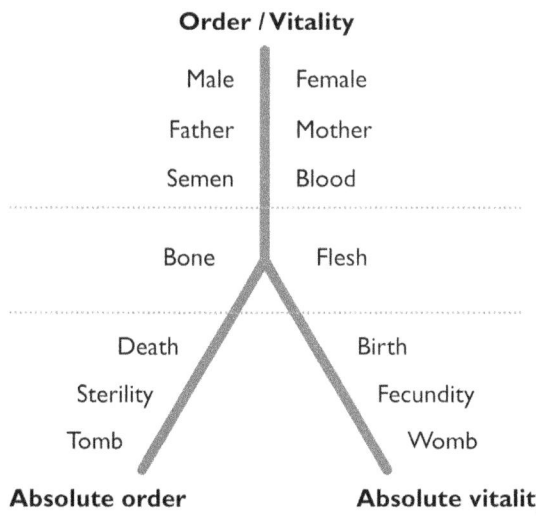

3.2 Bara concepts of the person (after Metcalf and Huntington 1991, Figure 7).

these conflicting experiences of time. The initial mourning and the intermediary period they would see as concerned with the polluting aspects of the decaying corpse, a contingent process which brings the linear experience of time into focus. By contrast, the final ceremony is interpreted as the point of regeneration and the time when the cyclical time of social order is re-instituted. A specific example of this interpretation is provided by Bloch's (1982) famous analysis of death rituals among the Merina of central Madagascar. The Merina response to the fact of death is characterised by displays of sorrow and what Bloch (1982, 214–215) refers to as 'self-deprecating' behaviour. The body is buried in an isolated grave and mourners, especially women, voluntarily associate themselves with the polluting evidences of the passage of linear time. By contrast, the reburial ceremony is conventionally associated with joy. The dry bones and dried remains of the flesh are exhumed and carried back to the ancestral territory of the dead person. The ceremony stresses the positive regrouping and return of vitality and life-forces to the ancestral lands. The body is placed in the appropriate ancestral megalithic tomb. However, existing bones of the ancestors are removed as part of this process, and the whole assemblage of skeletal remains are forcibly mixed and co-mingled in a complex ceremony before being returned to the burial chamber (Bloch 1982, 216–217). This act is interpreted by Bloch (1982, 217–218) as removing the anomalous individuality of the recently buried bones, associated as they are with the experience of the linear passage of time, and reintegrating them into the recurring blessings and authority of the ancestral community.

Thus, both Metcalf and Huntington (1991) and Bloch and Parry (1982) describe multi-stage funerary rites in a way which stresses the importance of time. I will return to this point in more detail in the next chapter. However, the importance of caves and tombs in these processes is that they are the material spaces in which the passage of time is experienced. Time acts upon the living, upon the decomposing corpse and upon the geologically active burial space. All of the Indonesian and Malagasy examples cited here have become extremely well known in the archaeological literature as potential ethnographic analogies for secondary burial rites (see for example, Parker Pearson 2003). However, caves have also been used for different kinds of long-term burial rite, which can also be analysed drawing on the insights of these two theoretical positions.

In southern Kenya in the late 1920s, Louis Leakey reported a former cave burial practice in the Taita Hills, Tsavo (Leakey quoted in Kitson 1931, 271–272). After death, Wataita people were buried in a shallow grave with a stone marking the position of the head.

After an intermediary period of between 1 and 2 years, the cranium was excavated and taken to a cave which acted as a family shrine. Leakey reports that the Wataita prayed and sacrificed to the bones as ancestors. These shrines were clearly numerous; Leakey was able to remove 120 crania from 12 shrines within a half mile radius of one village (Kitson 1931, 272). Ethno-archaeological fieldwork in Tsavo reported in Kusimba and Kusimba (2000, 18–20) provides a more recent account of the same practice. According to Wataita informants, the intermediary period burials in this region were placed under 1-metre-diameter stone cairns. After 2 years, the crania were then removed, as described in Leakey's account, to cranial display niches in nearby rock shelters, crevices or small caves. Kusimba and colleagues (2005, 247–250) recorded eight of these cranial display niches, two of which still contained skulls (Figure 3.3). Only married people with children were chosen to have their crania disinterred. Their accounts also provide much more detail on the meaning of the rite. Despite the partial nature of the remains, Wataita elders were able to relate the family relationship of each cranium over a five-generation period. The ancestors, in this case, were still individual beings. The Wataita sustained and placated their ancestors with gifts, left at the cranial display niche, of tobacco, meat and beer. In return, the ancestors protected the village from natural disasters, disease and witchcraft (Kusimba and Kusimba 2000, 21; Kusimba et al. 2005, 250). In this case, the passage of time during the intermediary period may have had the effect of concentrating the social relationships around the dead person into one particular part of the body, the dried and exhumed cranium. This one bone can then act on behalf of the deceased, as they enter into their new state as an ancestor. Interestingly, Leakey

3.3 A Wataita cranial display niche in Kajire rock shelter, Tsavo, recorded by Kusimba and colleagues (2005), which contained 308 crania (photograph by Chaprukha Kusimba).

reported (Kitson 1931, 271–272) that in cases where someone died by violence away from the Taita Hills, and it was not possible to retrieve the cranium, a limb bone could be recovered and used in its place. If this were not possible, a sheep's skull would then be substituted.

Multi-stage funerary rites may also involve mummification to deliberately slow down the effects of time on the body. Hertz (1960, 201) specifically included the actions of embalmers trying to control the decay of the corpse as an example of an intermediary period. In the central highlands of the Philippines, the final phase of multi-stage burial involves the natural mummification of bodies in wooden coffins (Picpican 2003). This rite has become well known in some locations, particularly Sagada, to the point where the burial caves and hanging coffins have become tourist destinations (Panchal and Cimacio 2016). Extended burial rites culminating in log-coffin burials have been a widespread practice on the island of Luzon, with Canilao (2012, 64–65) able to demonstrate through archaeological survey that the practice has taken place since the eighteenth century. Funerary rites in the entire highland region have been reviewed by Celino (1990). In the central Mountain Province region, which includes Sagada, the corpse is washed immediately after death. It is then dressed and seated for display on a specially constructed death chair for the duration of the wake. This is normally two or three days, but can be extended if close family members cannot arrive earlier (Celino 1990, 94–95). In other areas of the highlands, it is typically longer, usually around nine days. In her analysis of the beliefs around burial in the region, Celino (1990, 106–107) identifies the moment of death as the point at which the dead person becomes an ancestral spirit, one of the *anitos*. At this stage, the ancestral spirit is still in the world of the living; the purpose of the wake and the funeral procession is to aid them on their journey to the world of the *anitos*. At the end of the enthronement portion of the wake, the *anito* of the dead person is asked to intercede to ward off the chaos and evil believed to be present because of the occurrence of death. This effectively closes and truncates the intermediary period. The body is removed from the death chair for burial and bound in a foetal position within a burial blanket. The design on the blanket used to wrap the corpse is chosen so as to be recognisable to other ancestors of the same lineage. It is then carried in procession, along with a log coffin, by a group of close male kin to a burial cave. The burial procession can become rowdy, as it is highly propitious to have carried the body. Both the coffin and the bound body are rope-handled up the cliff face to a cave or alternatively, as at Echo Valley, Sagada, are suspended from the rock face. The bodies ultimately turn into mummy bundles owing to the dry air in the caves (Celino 1990, 98–103). The journey to the burial cave is thought of as being analogous to travel among the living; funeral processions start early in the morning because 'one starts travel early in the day' (Celino 1990, 107). Interestingly, the ethnographic accounts provided by Celino (1990, 102–103), seem to show that the journey of the *anito* to the spirit realm continues as the body mummifies in the burial cave. A series of staged feasts and rest days after the burial culminate in a final feast which recognises that the dead person is clearly in the spirit realm because they are demonstrably able to provide for the living as a benefactor and patron (Celino 1990, 103).

Hertz's (1960, 198–204) original characterisation of an 'intermediary period' in funerary rites is therefore a useful analytical tool for examining ethnographic examples of cave burial. The detailed ethnographies of death rites in Metcalf and Huntington (1991) and Bloch and Parry (1982) interpret the local meaning of the 'intermediary period' in substantially different ways. However, in all these cases, social and physical changes happen over time. Understanding the way that time is perceived is a major part of understanding extended funerary rites, and I will return to this topic in more detail in Chapter 4. One important way in which mourners would perceive time during the 'intermediary period' is by observing the physical decay of the corpse. As archaeologists, in order to use Hertz's (1960, 203) insight that the state of the decaying body would have been socially significant, we need to address the physical processes of decomposition and the post-depositional processes of taphonomy in more detail. We need to understand what the 'intermediary period' should look like in different kinds of archaeological deposit.

The taphonomy of human decomposition

Over the last 30 years, there has been extensive study of just this problem in both osteoarchaeology and forensic anthropology. Knüsel (2010) has reviewed the different ways that human skeletal remains have been studied in archaeology. There have been particular approaches to understanding past funerary rites from both an archaeological and an osteological perspective. In this paper, Knüsel (2010, 67–70) argues that two broad research traditions can be identified. One attempts to use osteological and archaeological information from death assemblages to reconstruction past lifeways, for example, population-level questions about human biology or using burial information to infer past social structures. Another is more focussed on interpreting and understanding circumstances of individual graves and bodies. This second tradition would unite aspects of the 'bioarchaeology' proposed by Buikstra and Beck (2006), the funerary archaeology discussed by Parker Pearson (2003) and the 'field anthropology' developed by Duday (2006). This research is obviously highly relevant to understanding what kind of traces an intermediary period should leave in the burial record. Knüsel (2010, 68–69) points to work by Henri Duday's former students which has attempted to use large archaeological cemetery data sets as long-term taphonomic experiments. This data has been used to infer a set of common sequences for bodily decomposition in different bodily orientations and grave types, as summarised by Knüsel (2014, 30–34). Similar data exists for other large mammal species, as synthesised by Morris (2011, 17–19). The use of a combination of large archaeological data sets and highly detailed excavation recording to interpret particular deposits can be referred to as 'palaeotaphonomy' (Quinney 2000, 12). The primary results of this research are synthesised in Knüsel (2014, 32), establishing a broad distinction between skeletal articulations which are 'labile', not supported except by soft tissue attachments and which therefore tend to disarticulate early without that support, and those which are 'persistent', with major ligament and tendon attachments and are therefore slower to disarticulate (see Figure 3.4).

The interpretation of bodily decomposition has also been approached though the 'neotaphonomic' (Quinney 2000, 12) method of drawing analogies from experimental decomposition studies. Particularly associated with research in forensic anthropology, this approach has been reviewed by Bristow and colleagues (2011, 280–284). The understanding of sequences of bodily decomposition gained from this research can be compared with those from palaeotaphonomic research. The pioneering syntheses of neotaphonomic research by Haglund and Sorg (1997, 2002) provide experimental data relevant to decomposition sequences (Roksandic 2002), bone weathering (Lyman and Fox 1997) and animal interactions with the body (Haglund 1992, 1997), all of which have been applied to archaeological material. Neotaphonomic research has the benefit of offering data on the duration and rate of decomposition processes in different experimentally observed situations. This, in turn, allows a more contextual and nuanced picture of decomposition in particular environments, which may explain different patterns of survival and movement for skeletal elements. For example, comparing the data from Haglund 1997 (see Figure 3.5 and Table 3.1) with the data from paleotaphonomic research in Figure 3.4 shows the effect of different environments and agents on bone survival, disarticulation sequence and survival of particular articulations. Neotaphonomic research also demonstrates that disarticulation which

Gestures and positions

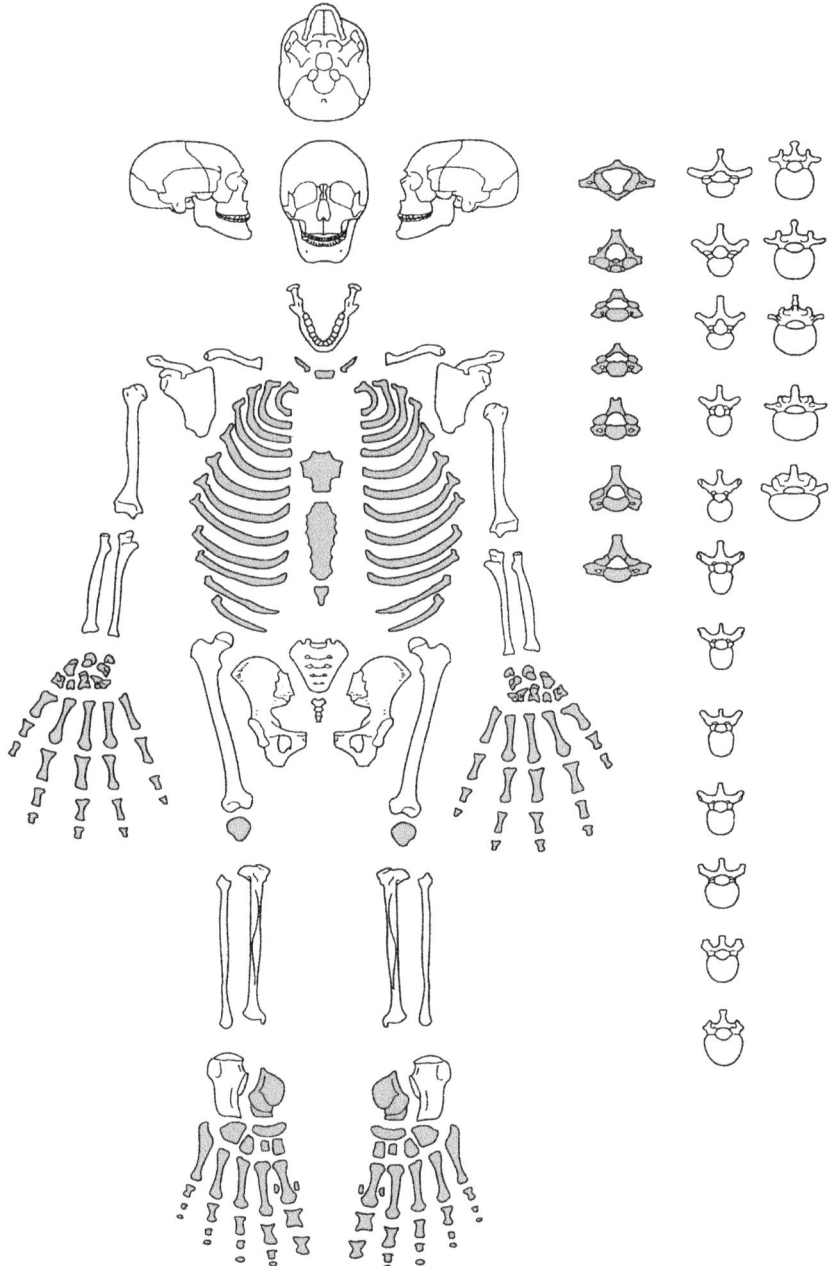

3.4 Sequences of bodily decomposition as suggested by palaeotaphonomic research. Shaded elements are those which have labile articulations and would be expected to disarticulate early (after Knüsel 2014, Figure 3).

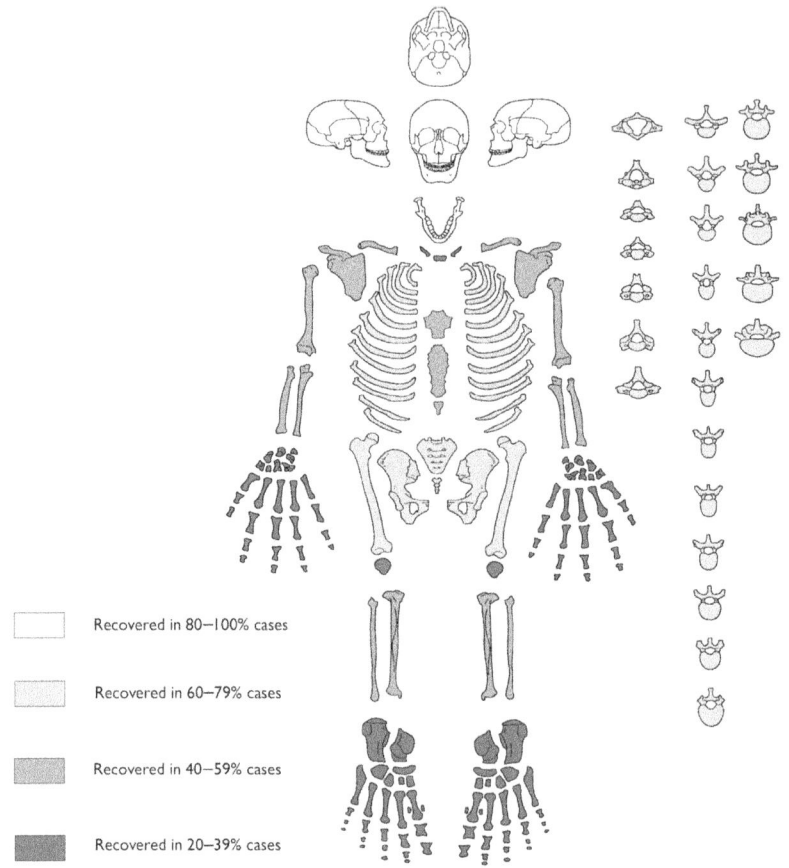

3.5 Frequency ranges for the recovery of skeletal elements scavenged by canids in the north-western United States based on 53 missing persons cases (after Haglund 1997, Figure 6).

involves animal or environmental agents can lead to the early separation of large bodily elements, which may then be preserved by other processes with continuing articulation of what are theoretically labile joints. An example of this process is the separation and preservation of articulated hand and foot elements of the hominin species *H. naledi* in the Rising Star cave system, South Africa (Dirks et al. 2015, 26–30).

Drawing on the results of both kinds of studies, it is possible to suggest a clear set of definitions and terminology to describe human remains from caves. These terms largely follow the suggestions made in Knüsel (2014), which is a valuable attempt to synthesise the Anglo-American osteoarchaeological research tradition with the 'field anthropology' advocated by French researchers.

Table 3.1 Time sequences for the canid-assisted disarticulation of human remains. Based on 37 examples from the north-western United States (Haglund 1997, Table 1).

Stage	Condition of Remains	Range of Observed Post-Mortem Interval
0	Early scavenging of soft tissue with no body unit removal	4 hours–14 days
1	Destruction of the ventral thorax accompanied by evisceration and removal of one or both upper extremities including scapulae and partial or complete clavicles	22 days–2.5 months
2	Lower extremities fully or partially removed	2–4.5 months
3	All skeletal elements disarticulated except for segments of the vertebral column	2–11 months
4	Total disarticulation with only cranium and other assorted skeletal elements or fragments recovered	5–52 months

Throughout this book, the term *primary burial* is used to describe a burial which is made and left undisturbed in its original location. This can include the burial of more than one individual if those burials take place simultaneously. The archaeological indications of primary burial include the complete preservation of the articulated skeleton with the survival of the labile articulations (see Figure 3.4) being a particularly diagnostic indicator (Knüsel 2014, 46). The important element in the funerary rites around primary burial is therefore the decision by the mourners to keep the body or bodies from being acted on by natural or cultural processes. Knüsel (2014, 47) also notes the historically documented phenomenon of 'delayed primary burial', which would apply to the very short intermediary period noted for the Philippine rites described here. In this case, bodily decomposition may begin in a shrouded or coffined body before burial, leading to the disarticulation of some of the labile connective tissue but the retention within the primary burial of all of the relevant bones.

Secondary burial is used in this book to refer to burials which included both a substantial intermediary period and more than one location in their associated funerary rites. I am following Knüsel (2014, 49–50) in giving primacy to Hertz's original conception of the intermediary period, so that, for the purposes of this book, a secondary burial is one which has been moved from one location to another over the course of the intermediary period. The Malagasy funerary rites studied by Metcalf and Huntington (1991) and Bloch

(1982) would both be classic ethnographic examples of a secondary burial in this sense. However, it is important to note that secondary burial may also include cases, such as the Kenyan example reported by Kusimba and colleagues (2005), where secondary burial is only given to part of the body. Osteologically, a secondary burial will almost certainly involve the complete disarticulation of the labile articulations. The absence of the relevant bones, such as the distal phalanges, has often been used as a marker of this kind of burial (Ashbee 1966, 37). However, the state of the persistent articulations in such burials will vary considerably depending on how long the intermediary period lasted. Neotaphonomic studies (Haglund 1997, Table 1) suggest that some persistent articulations could be expected to survive even in bodies that had been exposed to canid scavenging for up to 5 years. Where there has been active human intervention in the skeletonisation process, then this may leave evidence such as cut marks on bone (Knüsel and Outram 2006, 254–255). Other forms of skeletonisation where the body was exposed to animal actions will leave a variety of traces which have been well documented in neotaphonomic experimental data (Dirks et al. 2015, 17–19; Haglund 1992; Haglund et al. 1988).

Following the suggestion of Weiss-Krejci (2012, 125), I will use a separate term to describe burials where there is evidence of an intermediary period in the rite but no evidence that the body was moved from one burial site to another during this period. This type of funerary rite I will characterise as *successive inhumation*. The final deposit in a successive inhumation will very often consist of a comingled and collective assemblage of bones, but this appearance will have been produced by the disturbance and rearrangement of earlier bodies during the placement of later ones. This is the kind of rite suggested by Wysocki for Wayland's Smithy 1 (Whittle et al. 2007, 106–107 and see Figure 3.1). Historically and ethnographically, this kind of burial rite is well attested from the medieval period in Europe in church crypts and among Iroquoian groups in north-eastern North America (Knüsel 2014, 44). There are a number of osteological indicators which can be used to distinguish successive inhumation from secondary burial. By definition, successive inhumation will have taken place in a restricted area. If a reasonable sample of this area has been excavated, and once other taphonomic processes which may have biased the survival of certain skeletal elements have been taken into account, then all parts of the skeleton should be equally represented. In underground spaces such as caves and chambered cairns, the presence or absence of bone weathering is another important indicator of whether successive inhumation was being practiced. There is a considerable literature on the weathering of bone in a variety of environments (see Lyman and

Fox 1997 for a critical summary). The nature of the observed weathering on bone has been used to reconstruct burial practices in cave assemblages of various dates (Dirks et al. 2015, 22–24, for example). Understanding the nature of the space where successive inhumation has taken place also allows other inferences to be made, particularly from the presence or absence of bone modifications produced by scavenging animals. These are well understood at a species level from studies for a range of vertebrates (Haglund 1992; Haglund 1997) and invertebrates (Dirks et al. 2015, 24). Therefore, in cases where bones show these modifications, but the appropriate species would not have been able to access the burial site, they provide evidence for secondary burial as opposed to successive inhumation.

The final piece of terminology used in this book is the adoption of the term *multi-stage burial*. This describes deposits where it is likely that an intermediary period was part of the funerary rite but where there is insufficient evidence to distinguish whether the deposit should be thought of as a secondary burial or as the result of successive inhumation. This review of the potential traces of an intermediary period in the archaeological record has stressed the importance of understanding in detail the context and micro-environment of the burial. Where human remains have been deposited in caves, understanding cave processes will be vital to understanding funerary rites in caves. This will include the physical form of caves, the movement of artefacts and sediment within them and the behaviour of other animal species that use caves.

Cave processes

The active involvement of cave processes in human decomposition has been noted in forensic cases. Of particular interest here is Jama-Bezdan, Hrgar, Bosnia-Herzegovina, which was used as a mass grave in 1992 following the massacre of approximately 70 people (Simmons 2002). The site was excavated by a Physicians for Human Rights team in June 1997. It is an 80-metre-deep vertical shaft cave which terminates in a small, 4 × 7-metre chamber. Prior to its use as a mass grave, the chamber had been used as an informal dump, and the chamber contained a talus cone of debris centred beneath the vertical shaft. As the bodies dropped onto the top of this talus cone, they were affected by a range of processes directly connected to the cave environment. Three properties in particular influenced the way in which the bodies were transformed in the cave. The unstable talus slope caused the first bodies deposited to move some way down slope as they were dropped into the chamber, where they remained in a more or less articulated state (Simmons 2002, 267). The moist cave environment and high numbers of invertebrates accelerated decomposition (as discussed by Simmons et al. 2010, 891–892), as did the elevated temperatures associated with large numbers of decomposing bodies. Therefore, some bodies decomposed rapidly high on the talus slope, probably trapped by the build-up of the first bodies into the chamber. These bodies rapidly became highly fragmented, and skeletal elements, especially crania, were moved by water and gravity into a co-mingled deposit at the base of the talus slope (Simmons 2002, 267–268 and see Figure 3.6).

This example indicates the need for a clear understanding of geomorphology, which is the study of the processes which governed the geological formation of caves and therefore their shape. It also indicates the need to understand the way in which water and sediments move in caves. Most of the sites with human remains from Britain (see Appendix 1 and Figure 1.1) are karstic caves, formed by the slow dissolution of limestone by water. This gives them a particular set of morphological characteristics. There are also a small number of burial caves, particularly in western Scotland, which have formed through coastal erosion of other rock types (Bonsall et al. 2012, 11–13). Additionally, caves can form in limestone through a process of mass movement. These caves are usually vertical fissures created by the slippage of large blocks of limestone. The Ryedale Windypits in north-east Yorkshire are examples of caves of this type which contain

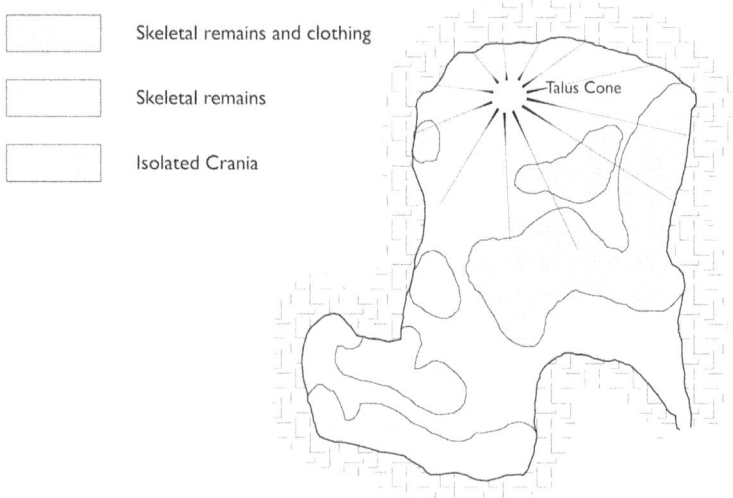

3.6 Plan of the recorded position of human remains at Jama-Bezdan, Hrgar (after Simmons 2002, Figure 13.3).

human remains (Cooper et al. 1976). Generally, cave formation processes in limestone are well understood; Jennings (1985) provides an accessible overview of this literature. From the point of view of this study, karst geomorphology provides a clear set of descriptive terms for the common features of limestone caves.

One important distinction in understanding how a cave has formed is between *phreatic* and *vadose* environments. Phreatic caves were formed beneath the water table, so that the whole developing system was entirely filled with water. By contrast, vadose caves were formed above the water table, and therefore the developing system would have contained both an air space and an active underground river (Weight 2002, 627). This difference has important consequences for the form of the resulting cave, or cave section. Phreatic caves are typically rounded or elliptical in cross-section, as the water can dissolve all surfaces simultaneously (see Figure 3.7). They often have areas of distinctive scalloped erosion, caused by turbulence in the rapid high-pressure flow that created them (Jennings 1985, 144–149).

Vadose water flow produces a much wider range of cave forms. Underground streams cut into the bases of existing phreatic tubes, widening and deepening them. Joints and fissures within the limestone bedding can be enlarged, and the whole system will tend to develop a branching network of tributary cave streams (Weight 2002, 629). This can lead to the formation of complex 'maze caves', where large numbers of intercutting passages and chambers link together (Fairchild

3.7 Phreatic portion of Fairy Holes Cave, Whitewell, Lancashire. Note the rounded cross-section and scalloped erosion on the left side of the cave wall.

and Baker 2012, 57–58). Vadose parts of the system are typically larger and, being exposed to the air, they are also subject to other weathering processes which can lead to expansion of the cave through roof collapse (see Figure 3.8). Stream flows in caves are often both rapid and highly erosive. Archaeological materials within cave sediments are therefore extremely likely to be transported in streamways. For example, crania from Romano-British burials within chamber 4 at Wookey Hole, Somerset, appear to have floated in the streamway of the river Axe to new locations near to the cave entrance (Hawkes et al. 1978, 25–29). It is also possible to see cases where this property of caves has been deliberately exploited, such as in the formal deposition of Neolithic human remains and Bronze Age metalwork in the underground passage of the river Lesse in Le Trou de Han, Namur (Warmenbol 2014, 69–73).

Cave roof collapse is one of the formation processes which leads to the creation of *dolines*, vertical shafts into caves. However, there a range of other erosional processes which will created a doline from above (Jennings 1985, 106–113). Solution associated with localised weaknesses in the limestone will produce a vertical shaft which tends to open out at depth (see Figure 3.9).

Once air enters a forming cave system, the wider range of erosional processes and slower stream flows leads to the build-up of both

3.8 A variety of vadose erosional processes are visible in the entrance chamber of Dunald Mill Hole, Lancashire. These include the down-cutting of the floor by the streamway, which now flows beneath the boulders in the foreground of the picture, and substantial roof collapse caused by aerial weathering.

cave sediments and the deposition of flowstones (Jennings 1985, 152). In the outer zones of caves, where most of the evidence for Neolithic human remains is found, these processes provide much of the physical environment. *Speleothem*, which is the collective scientific term for stalagmites, stalactites, flowstones and other structures formed from the precipitation of calcium carbonate within the groundwater, only forms under certain conditions. It forms in caves which are above the water table and which have a supply of groundwater and a circulation of air to remove the carbon dioxide waste products from precipitation (Fairchild and Baker 2012, 7). Stalagmites and stalactites will not form where the air is too turbulent, and hence they do not usually form in the daylight zone of caves, although other forms of speleothem, particularly travertine and tufa, will do so (Mourne et al. 2012, 63 and see Figure 3.10).

3.9 A solutional vertical shaft into the chamber at Heaning Wood Bone Cave, Cumbria.

3.10 Granular tufa deposit forming at the back of the Cave Ha 3 rock shelter, North Yorkshire.

Speleothems can be shown to have interacted with human bone assemblages in caves in a number of different ways. In some cases, human bone is reported as having become cemented into flowstones, for example at Carsington Pasture Cave, Derbyshire (Barnatt and Edmonds 2002, 117). Evidence for the deliberate use of speleothem formation in the Italian Neolithic comes from the lower chamber at Grotta Scaloria, where pottery vessels were located on flowstone surfaces, where they collected the precipitating water from stalactites and became petrified parts of new stalagmites (Whitehouse 2015, 57–58). Leach (2008, 51) also noted a link between Neolithic cave burial in Yorkshire and the active deposition of tufa. Tufa forms in the presence of micro-organisms and is generally deposited in active streamways around accumulated organic material, although it can also form in the daylight zone of caves (Mourne et al. 2012, 63). It can vary in texture from extremely dense and laminated to porous and granular (Dabkowski 2014, 72). Tufa was clearly an auspicious material in some non-cave archaeological contexts during the Neolithic. Davies and Lewis (2004, 8) report the deposition of compressed balls of tufa in small Late Mesolithic or Early Neolithic pits at Langley's Lane, Somerset. An Early Neolithic burial from Prestatyn, North Wales (Schulting and Gonzalez 2008, 303), was closely associated with both a wide area of Late Mesolithic cockle middens and an extensive area of tufa deposition. Bateman (1861, 89–90) also reports an example of a burial in a tufa deposit associated with faunal remains from Monsal Dale in Derbyshire.

The aerial weathering processes which affect the outer zones of caves also commonly lead to the creation of limestone scree deposits within and outside caves. Screes can be classified as either clast-supported, where the angular limestone fragments which make up the deposit are all in direct contact with one another; or matrix-supported, in which the fragments are separated by finer sediment particles. The presence of archaeological material in these screes is often used by cave geologists to indicate that these are recent deposits which are still forming (Waltham and Murphy 2013, 138). Artefacts and human remains which have been placed on the surface of clast-supported screes are highly likely to be displaced downwards through the airspaces in the deposit. In practice, even within matrix-supported screes, there is a high probability that small dense artefacts and bones will have moved downwards. For example, Early Neolithic human remains from the same area of George Rock Shelter, Vale of Glamorgan, which were probably deposited together on the same scree surface, were displaced over 1 metre of the vertical stratigraphy (Peterson 2013, 270)

Breccias are extremely mixed sedimentary deposits which have formed in place within caves. They typically contain large limestone clasts within a clay matrix with active speleothem formation which acts to cement the material together. The different materials within a breccia are usually brought together by the mass movement of deposits through some type of debris flow event (Jennings 1985, 165). Breccias have been an important area of study for cave palaeontology and the cave archaeology of earlier periods. The debris flow events which created breccia layers within Pontnewydd Cave, North Wales, were highly erosive. They transported both artefacts and hominin teeth from parts of the cave system which are now destroyed and emplaced them in breccia deposits considerably further into the system (Mourne et al. 2012, 61–62). The mechanisms and power of debris flows within archaeological caves in particular have been reviewed by Mourne and colleagues (2012, 61–63) and by Collcutt (1984, 54–59). The Neolithic human bone from Cattedown Cave, Devon, was discovered, along with deer, wolf and hyaena bones, in a heavily cemented breccia deposit, presumably having been transported by a debris flow which had also accumulated Pleistocene material (Worth 1887, 109–111).

Smaller particles may also be moved into and around caves by both water and airflow. Cave sediment formation processes generally are reviewed by Ford (2001) with particular reference to British caves. Wind-borne sediments are usually a minor part of most cave deposits (Ford 2001, 17), although wind-blown sand is a major component of the deposits within coastal rock shelters such as An Corran, Skye (Saville et al. 2012). Fine sediment deposits in caves are largely the result of water transport; the precise nature of the particles in these sediments depends on the external source from which it is being transported (Figure 3.11). However, the principal influence on the type of deposit in the cave itself is the speed of the water flow as it was being deposited. Under relatively rapid flow, coarser particles will be preferentially deposited. When cave passages are partially blocked by screes or roof falls, finer silt and clay particles will be deposited (Ford 2001, 10–14). The major influence that this kind of cave sediment has on archaeological deposits is to bury artefacts and bodies that were previously exposed on a surface.

Cave formation and sedimentation processes can therefore be seen as more than a mere backdrop for cultural practices. Burial rites in caves would have taken place within environments which would have had considerable effects on the human remains. We also need to be wary of treating the information we have about taphonomic and geological processes simply as a barrier to our understanding of funerary rites which needs to be overcome. I will explore this topic in more detail in Chapter 4, but, as this chapter has shown, the active nature of decomposing bodies and cave systems would have formed the material context for the intermediary period in multi-stage burial rites.

3.11 Bands of alluvial silts and clays in section in Temple Cave, Whitewell, Lancashire.

Conclusions

This review of the way that human action, bodily decomposition and cave processes come together in the intermediary period begs a larger question. What was the purpose of the intermediary period in cave burial? There are a number of major themes around the interpretation of ritual which, as I have discussed here, may lie behind the development of this kind of burial. Funerary ritual has an important relationship to memory and specifically to the management of the memory of the deceased; for example, see Fowler's (2003) analysis of the use of decay and fragmentation in the Early Neolithic of southern Britain for this purpose. It is also the case, as can be seen from the ethnographic review in Metcalf and Huntington's (1991) work, that it is possible to interpret multi-stage burial as an attempt to manage the transformations inherent in death; funerals can be seen as rites of passage. Alternatively, it is possible to regard multi-stage burial more in the light of a social tool. Bloch's (1982) analysis of the funerary rites of the Merina can be considered in this light. Ritual, in this case, can be seen as a communicative social tool which is used to achieve certain social ends. I believe that what all of these accounts have in common is that they draw upon the agency of death. Regardless of the precise way in which bodies, things and the environment work together to create the human experience of the world, death as an event disrupts this process. It is both a spur to action – bodies start to decompose and social obligations go unfulfilled – and it fundamentally changes the existing structuring conditions through which those actions make sense.

I think that it is perhaps more helpful to think of the intermediary period not as something which is imposed on bodies and things by external social norms but rather as something which arises out of the way that people and things act during death. This directly contradicts one aspect of Hertz's original characterisation of the intermediary period. He argued that it was primarily a social phenomenon, which was why it was applied more noticeably when influential people died and was not applied at all in most infant burials. However, this characterisation depended on a view of social being as something separated from and 'grafted onto' a physical body (Hertz 1960, 207). Instead, I would argue, it is more useful to imaging social agency as an embodied, material phenomena drawing on the interactions between bodies, places and objects. The ability to act in a way which is meaningful to living observers is not something which is intrinsic to either living people, dead bodies or caves. Rather, actions are perceived as taking place because of the network of interactions between all these things.

This broad idea has become increasingly influential within archaeology over the last 15 years in a variety of different theoretical approaches, such as symmetrical archaeology (Shanks 2007), relational realism (Fowler 2013), assemblage theory (Robinson 2017), embodied 'affects' (Mlekuž 2011) and 'new materialism' (Conneller 2010), all of which have been grouped together as examples of the 'post-humanist' turn in archaeological thinking (Harris and Cipolla 2017, 129–149). I will review the origins and connections between these ideas in much more detail in Chapter 4.

However, in the specific case of how the intermediary period may develop from the way that people and things are affected by death, the important connection between the material world and the living mourners is the perception of the passage of time. In particular, the relationships which make up this network of interactions are perceived by the living by observing the physical clues which show them that time has passed. Therefore, when a death occurs, these networks are disrupted and have to be re-formulated. As these networks have built up through time, then long-standing networks would tend to be more complex than recently established ones. This is the explanation for Hertz's observation that the intermediary period is not applied in most cases for infant burials. The length of time a network has been existence influences the scale of social disruption felt when a death takes place. The transitions in bodies, places and obligations still take place, but the more complex networks will require more obvious manifestations of the process. What we identify as the intermediary period arises from the way that the body, the social obligations of the dead and the living and the cave are reformulated by the fact of a death.

4

How do caves act?

In the last chapter, I suggested that the intermediary period around death was something which involved the funerary rites carried out by living people but that it also involved the biological processes of bodily decay and the geological processes which go on within all caves. Any funerary rite where the body was placed in a cave would have ensured that the body was being acted upon by all these factors. Archaeologists studying funerary rites often treat taphonomic processes, such as bodily decay and cave sedimentation, as things which hinder our understanding. To adopt Michael Schiffer's (1976, 11–12) terminology, they are the natural formation processes which need to be understood so that we are able to connect the archaeological record with past human activities. For example, Zemour (2011, 258–259) discusses whether Early Neolithic burials of south-western France were deliberately buried in a flexed position and on their sides. She concludes that possible taphonomic changes in some bodies and the nature of the cave space they are buried in makes it impossible to identify any such cultural choice in the burial rite. Patrick (1985) has described the conventional understanding of archaeological research as following a metaphor of 'the record'. Human social action in the past is understood to have created a record. Archaeologists uncover fragmentary remains of this record and must 'strip away' the distortions created by everything that has subsequently happened to the objects they discovered. Once they have done this, it is assumed that the past social understandings which created the record can be reconstructed. Patrick (1985) and Barrett (2001) have both criticised this vision of the aim of archaeological research as unrealistic. Barrett (1988, 10–12) in particular suggested that, rather than being regarded as a record to be transcribed, archaeological material should be regarded as objects which were actively used by past people to create their social structures. When Neolithic people chose to use caves for funerary rites, they would have done so with a clear understanding of both how bodies decompose and how cave environments would influence that process. The taphonomy and geology are not accidents which conspire to prevent us understanding Neolithic cave burial properly; instead they would have been active and meaningful contributors to that funerary rite.

Previous studies in cave archaeology have suggested that cave systems can 'act' on people in different ways. In the Slovenian Neolithic example discussed in Chapter 1, Mlekuž (2011) argued that the bodies of sheep and shepherds were created through the repeated use of caves as shelters. The cave walls therefore become an important part of the creation of a bodily identity connected with Neolithic domesticity and pastoralism. In this case, caves were assumed to have the ability to act on living bodies. As discussed in Chapter 3, Leach (2008, 39) has suggested that caves, and particular tufa deposition, acted on dead bodies. At Cave Ha 3, North Yorkshire, four individuals were buried within an actively forming tufa deposit while their bodies were still articulated. Leach (2008, 51) argues that the petrifying properties of tufa springs were actively incorporated into the burial rites at this cave, deliberately invoking the agency of the cave system. This type of research leads to a wider range of questions about the agency of caves and bodies. What exactly do we mean when we talk about a cave having agency? Is *agency* an appropriate term to use to describe something inanimate like a rock formation? The wider question of how people and the world interact over time has been of increasing importance within prehistoric archaeology over the last 15 years. The connections between living people, dead bodies, artefacts and landscapes are all central to what has been described (Harris and Cipolla 2017, 129–149) as the 'post-humanist' turn within archaeology. In the first section of this chapter, I will review how some of the wider social theory about agency has been developed and applied in archaeology. I have discussed this literature in more detail in a recent publication (Peterson 2018), and therefore some of this chapter summarises arguments I have already explored. However, to address the questions established at the end of Chapter 3, it will also be necessary to explore the connections between caves, bodies and material culture through time. Therefore, the second part of this chapter draws on a wider literature about the material and embodied nature of memory and time.

What do we mean by *agency*?

Theory about agency was introduced into archaeology in an influential paper by Barrett (1988). In this paper, he attempted to shift archaeological analysis away from studying patterns in artefacts to finding a methodology for thinking about the way that relationships between people were structured (Barrett 1988, 8–10). To do this, Barrett drew to a large degree on the 'structuration theory' of Anthony Giddens (1979, 1984). Structuration theory is, in many ways, a classic example of the problem that I want to address in this chapter. It provides a holistic model of social institutions as they are constructed in specific human actions (Giddens 1984, 34). Therefore, it should be helpful to archaeologists trying to understand broader social issues from detailed evidence about particular human actions (Barrett 1988, 8). However, the sociological data used by Giddens to develop his argument is very different to the embodied material evidence we encounter in archaeology.

Individual human agency, as described by Giddens (1979, 56, 1984, 5), moves through three stages. First, there is the motivation for the action; then, there is the rationalisation of the action; and finally, there is the reflexive monitoring of the action. However, these stages take place within a surrounding structure made up of the existing conditions within which the action takes place and its unintended consequences. This surrounding structure motives the action and provides the context for its rationalisation and the comparative standard which allows it to be reflexively monitored. This model provides a theoretical methodology for working out a recursive relationship between individuals' thoughts and actions and the social structures around them, the duality of structure. The central contribution of structuration theory to an archaeological analysis of agency is the way that this duality of structure uses memory and the experience of time to connect human action, bodily experience and social institutions (Giddens 1984, 25–26). However, archaeological writers have also perceived a number of areas where Giddens' work requires elaboration to fit with archaeological concerns and evidence. Barrett (1988, 27) was critical of a lack of engagement with the material world. Similarly, Gardner (2004, 7) suggested that problems of subordination and domination needed a more in-depth analysis. Both these writers adopted elements of the work of Pierre Bourdieu (1977) on the daily practice of everyday life into their analysis to address these concerns.

The relevant parts of Bourdieu's (1977, 1990) work are concerned with developing a theory of practice for the study of society. For archaeologists, his most influential idea has been the concept of 'habitus', the analysis of daily routines of everyday life. Habitus is 'knowing how to go on': the unconscious knowledge of what constitutes appropriate behaviour used by people to get through their day-to-day life. As such, it is generally not consciously articulated and is very variable between different cultures (Bourdieu 1977, 72). Social structures and institutions constrain the actions of habitus. However, they are also created from and reinforced by the actions of habitus (Bourdieu 1984, 170). Bourdieu uses the action of memory to overcome the apparent circularity of this argument. He (Bourdieu 1977, 87) discusses the concrete example of the way in which the memory of learning within the family underpins the way learning is experienced in school, which in turn creates new memories which underpin the way learning is experienced in later life. For archaeologists, one positive result of Bourdieu's focus on theory as practice is that it provides a description of agency which is closely linked to material objects. In his detailed examples, the structures which are developed in and from habitus are concrete physical things. Relationships between people are mediated through objects and architecture. For example, one of the ways in which he analysed the differences in social norms between different classes in France was to look at the unspoken practices around social dining (Bourdieu 1984, 193–200). To do this, he examined the contrasting expectations of each class for how people would speak and behave, what they would wear, the kind of food that would be prepared and how it would be presented. In essence, the habitus of social class was presented as something that was articulated through bodies, food and material culture. Figure 4.1 shows the results for one part of that analysis, examining choices about which kinds of tableware were thought appropriate for each broad social class.

Returning to my original question of how caves, bodies and objects act, Bourdieu's theories of practice represent an important further step and shows that habitus is expressed through bodies and material culture. This is much more helpful in interpreting the way

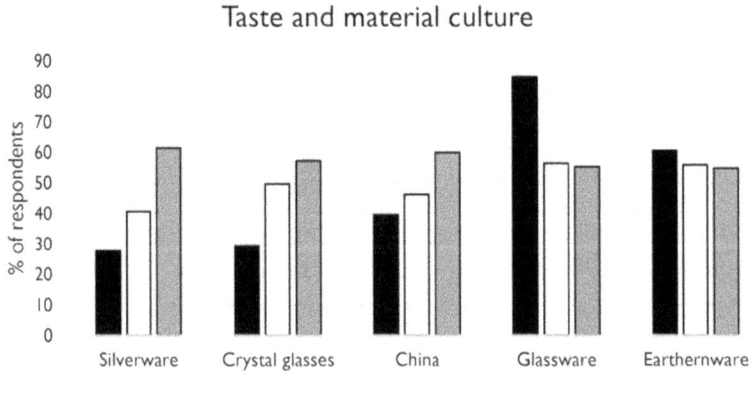

4.1 The material culture of class as expressed in choices of tableware for social dining in France. Based on data in Bourdieu (1984, Table 19).

that living people may have interacted with caves. However, even in this case, agency is primarily something which people have. They may express it through material culture, but the material culture itself does not act. The examples in Chapter 3 show that caves and bodies were active in a different kind of way; they are not merely expressions of the agency of living people. Fortunately, there are other approaches to how people and things interact which address the agency of non-human actors more directly.

Ingold (2000, 172–173) discusses the traditional distinction in anthropology between the cultural world and the natural environment. In particular, he focusses on the different ways in which people and animals are thought of as interacting with the environment. Conventionally, people are thought of as inhabiting a 'social domain' characterised by intentional motivations, while animals inhabit an 'ecological domain' characterised by adaptive responses to the environment. Ingold (2000, 174–181) overcomes this distinction by looking at how social and ecological explanations have been applied to the structures made by people and animals. The structures that animals make are usually described as being created when the animal's biological imperatives come up against a particular set of ecological conditions. When humans build, by contrast, this is assumed to be the result of intentional human design. Ingold (2000, 181–186) draws on the work of Martin Heidegger (1971, 145–161) to create a 'dwelling perspective' which can be used to look at the creation of both kinds of structure in the same way.

Once again, the passage of time is a key element of the argument. Ingold (2000, 187–188) argues that both human and non-human animals inhabit a world which is already structured. They respond to the buildings and environment that are already around them as they build new structures. These structures in turn become part of the environment which is responded to when future building occurs. This can be seen in the way that beaver dams and lodges both modify the environment and respond to earlier modifications. A prehistoric example of this process can be seen in the Late Mesolithic deposits at Stainton West, Cumbria. Excavated evidence in the palaeochannel here has been interpreted as the remains of a beaver lodge and dam (see Figure 4.2). These structures modified the local environment, producing clearings, fords and ponds, in a way that also made the landscape particularly attractive to human settlement. On the gravel islands next to the river, there is evidence for substantial Mesolithic settlement, which was probably sited to take advantage of these changes (Brown and Clark 2011, 100–104). In this case, both people and animals would have been responding to transformations in the environment

4.2 Probable beaver lodge in the Mesolithic organic deposit within the former course of the river Eden at Stainton West, Cumbria. Photograph by Fraser Brown/ Oxford Archaeology (North).

created by one another over time. The large fallen tree in the centre of the lodge was ring-barked by human hunters clearing the area around the river. The beavers then used this as the focal point for their lodge. There was subsequently evidence of human wood-working on the top of the lodge (Fraser Brown, pers. comm.)

Ingold (1993) has elaborated elsewhere on the importance of the experience of time in understanding the dwelling perspective. In this work, he developed the term *taskscape* to describe a group of related activities, analogous to the way that a landscape is an array of related physical features (Ingold 1993, 158). Taskscape can be thought of as a material manifestation of the kind of structures discussed by Giddens and Bourdieu; it is both the medium within which actions take place and is created from the results of those actions. Taskscapes also depend on the experience of time passing and, importantly, Ingold (1993, 159) also finds a way to describe the passage of time in an embodied way. He uses the term *temporality* to describe a conception of time which is not calibrated to an external constant but is instead derived from the experience of doing the activities in the taskscape. When people or animals do things, they make time pass. Temporality is the time of the participant.

Ingold's work develops both of the broad themes around the study of agency I have highlighted so far. Firstly, some version of what Giddens describes as the duality of structure is helpful to understanding agency; actions develop within structures and structures are created from actions. Secondly, time and memory are a central part of how the duality of structure operates. What is novel in Ingold's approach is that he extends Bourdieu's interest in objects and space to provide a description of both the duality of structure and the passage of time which is rooted in the material world and bodily experience. This recasting of the argument in terms of temporality and taskscape also allows Ingold to extend the definition of agency beyond those animals that have human consciousness and intentionality.

There is a thorough review from an archaeological point of view of the literature on the agency of inanimate objects by Alberti and Bray (2009). They point out (Alberti and Bray 2009, 339–340) that archaeological approaches to the problem of things that act have generally followed the work of either Alfred Gell (1998) or Bruno Latour (2005). The difference between these two approaches is largely that Gell makes a distinction between the 'real' agency of living subjects and the ascribed agency of passive objects. Latour, on the other hand, works with a 'flat ontology' which would describe the agency of people, animals and things in the same way. Gell (1998, 16) sets out by defining agency in a way which ties it strongly to deliberate human intentions. Despite this definition, he then develops an argument which suggests that art objects have an extremely powerful kind of agency (Gell 1998, 13–17). To do this, he introduces the concept of the 'index'. An index is any object which allows people to make a 'causal inference' Gell (1998, 13–15); it is anything which allows the viewer to infer that an active agent created the object. Gell (1998, 20–21) suggests we divide agents into two groups. Humans acting intentionally would be classified as 'primary agents'. These are contrasted with 'secondary agents', which are the objects that primary agents use to distribute their agency. These secondary agents are not in any sense less authentic. They are the dispersed material manifestations of primary agency (Gell 1998, 140–141). For Gell, objects have agency as distributed parts of the people who have made and used them. Any other person who encounters such an object is able to make inferences about the primary agent. The object 'embodies intentionalities' (Gell 1996, 36).

The way in which Gell imagines distributed agency working is illustrated by his analysis of the different claims made of non-western 'artefacts' and 'artworks' in the modern art world (Gell 1996). Traps, in a particular part of Gell's argument, are an excellent example of how primary agency can operate at a distance and over time. They index both the trapper's intentions and their knowledge of how to subvert the habitual behaviour of the prey animal to catch them. Discussing the specific case of the eel traps made by Ankave people in New Guinea (Figure 4.3), Gell (1996, 32–34) suggests that the traps also draw on Ankave ideas about the role of eels in their cosmology and mythology. They are constructed in a particularly complex and elaborate way to index the power of the eels. Therefore, an Ankave eel trap indexes knowledge through time, recalling the skills required to construct it and ancestral beliefs around eels.

Gell's description of the agency of inanimate objects is very persuasive in its own terms. However, this is not a completely helpful solution to the problem for prehistorians. It does not provide a straightforward methodology to identify what it was about an object which may have led people in the past to ascribe agency to it. Gell's secondary object agency is an example of what Pels (1998, 94) would describe as 'animist' object agency, in which things have life because an external soul or spirit is perceived as animating them. Pels (1998, 95) contrasts this with 'fetishist' object agency, in which an object's power comes from the very nature of the materials of which it is comprised. This sort of materialist and embodied perspective on object agency has been considered in archaeology and in the wider social sciences by a number of different writers and will inform the analysis in this book. Examples of this kind of thinking include the 'relational realist' archaeology proposed by Fowler (2013, 20–67), Tim Ingold's (2011) 'meshworks', the 'symmetrical archaeology' described in Shanks (2007) and the 'assemblage theory' associated with De Landa (2006). All of these approaches to object agency have three things in common. These are a fundamental critique of the distinction between active subjects and passive objects; a focus on the relationships between objects and people (whether these are described as *networks*, *meshworks* or *assemblages*); and a 'flat ontology' which does not prioritise one kind of agent or structure over another.

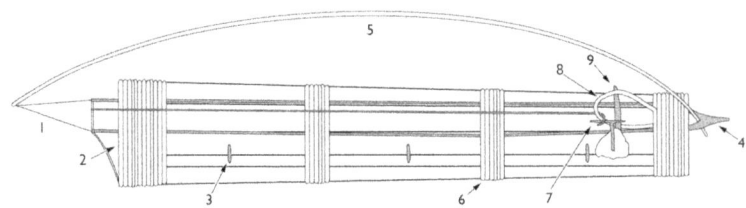

4.3 Ankave bark and rattan eel trap: 1) rattan, 2) door, 3) wooden pin, 4) forked piece, 5) spring, 6) spiral, 7) small stick, 8) rattan loop, 9) T-shaped trigger onto which frog bait is tied (after Lemonnier 2012, Figure 15).

These three themes all emerge to a greater or lesser extent from the 'Actor-Network Theory' developed by Latour (2005). Latour suggests a contrasting way of thinking about the agency of inanimate objects to that developed by Gell. Central to this analysis is a critique of any distinction between humans as active subjects and things as passive objects (Latour 2005, 70–74). Whereas Gell (1998, 20–21) divided primary human agents from secondary material agents, Latour (2005, 46) declared 'an actor is what is made to act by many others'. He develops the term *actant* to describe the property of making a difference. People, animals or objects are all, in Latour's analysis, equally capable of making a difference to any given situation. The 'actant' is introduced in a way which does not require it to possess any kind of consciousness or intentionality:

> Kettles 'boil' water, knives 'cut' meat, baskets 'hold' provisions ... if action is limited a priori to what 'intentional', 'meaningful' humans do it is hard to see how a hammer could act. ... any thing that does modify a state of affairs by making a difference is an actor. ... the question to ask about any agent is simply the following: Does it make a difference in the course of some other agent's action or not? (Latour 2005, 71)

If we think of objects as actants, we can see how they allow, afford, permit, encourage, block or forbid actions (Latour 2005, 72). Objects, in this argument, don't themselves 'have agency'. However, there is also no separate category of people with intentions who are the primary agents. Instead, agency is thought to exist in the network of relationships between all the actants. These relationships between actants are the key part of actor-network theory. Despite the critique which Latour offers of the way the distinction between subjects and objects has been analysed, it is important to note that he does maintain that there is something significant which separates people from things. Actor-network theory is a tool for the analysis of human societies and institutions; therefore, the presence of human beings is necessary for that analysis to be meaningful (Latour 2005, 78). The 'flat ontology' and the critique of the subject/object distinction are tools which must be in place before the analysis can be carried out properly, but if the network does not involve any people, the analysis is not relevant (Latour 2005, 75–76). There has also been a debate, helpfully summarised by Fowler and Harris (2015, 129–133), about whether Latour places too much emphasis on the objects within a network (a position which Ingold [2011] has adopted) or too much emphasis on the relationships between them (Harman 2009). Fowler and Harris

(2015, 133) argue that to resolve this debate it is necessary to accept that both the objects and the relationships in the network are equally real; hence their espousal of a 'relational-realist' archaeology.

The passage of time is also a central analytical concern of actor-network theory. Specifically, Latour (1999, 69–79) addresses the issue of change and time through the concept of the 'circulating reference'. He describes phenomena being studied by scientists as circulating along a chain of connections which link the different temporal incarnations of the phenomenon together. Every time a phenomenon is described or analysed, it gains some properties and loses others. However, not all the properties change at the same time, and therefore the network of relationships persists. They can be thought of, in similar ways to Gell's indices, as providing the frame of reference that allows the passage of time to be observed. They connect the present network of actants with all previous networks that they may have been a part of (Latour 2005, 200–201). Therefore, caves and bodies would have been part of the network of connections created as people made use of them. As the reviews of taphonomy and sedimentary processes in Chapter 3 show, both bodies and caves would have provided compelling indications of the passage of time. As physical and temporal actants, they would have linked older and more recent networks together, helping to provide the structure that made sense of the relationships.

Ian Hodder (2012) has also provided an interesting holistic account of the archaeological connections between people and things. His description of this process is one of 'entanglement'. He would see the relationship between things and people as one of dependence and dependency (Hodder 2012, 206–207), which might be characterised as positive and negative consequences of the way people and things interact. *Dependence* describes the enabling consequences of the relationship. Hodder (2012, 44) uses the example of the network of social and technical relationships involved in making a fire. Tools, flammable materials and sources of ignition depend on people to bring them together, and people depend on them to achieve warmth, cooking or sociability. All of these dependences are a part of the fire, and no particular one has priority. *Dependency*, by contrast, describes the way that objects and people are limited by these relationships. Hodder (2012, 57) gives the example of the constraining effect of raw material qualities on the types of stone tools which can be manufactured. *Entanglement* describes the way that these two relationships interact. Hodder (2012, 88–112) choses this term instead of Latour's *network* to give a sense of the directionality of connections implied by the constraints of dependency. Hodder (2012, 88–89) talks of a dialectic between the way that dependence allows people and things to rely on each other; and dependency, when limits of resources, techniques or social structures are reached and new dependences arise within the constraints set by the – now no longer viable – previous dependences. Hodder explicitly does not have a 'flat ontology' in his analysis (Harris and Cipolla 2017, 105). The relationship between people and things is at the centre of his analysis, as is the relationship between things and the experience of time. This is a point I will return to in the second part of this chapter.

Several common threads can be drawn from this review of ideas around agency. I want to pull some of these together to develop a theory of practice which allows us to combine the study of human agency, the agency of bodies and the agency of caves. The first important common principle is that agency and structure are recursive. The concept of the duality of structure proposed by Giddens (1979, 69) is echoed by Bourdieu's (1984, 170) structuring and structured structures. The environmental responses which form Ingold's (2000, 187–188) dwelling perspective are similarly recursive, as is the way that Gell (1996, 36) supposes that inferences can be drawn from an index and the way that Latour (2005, 71) sees that actants 'make a difference' to other actants in a network. The second common thread that I wish to emphasise is around the relationship between agency and intentionality. Even Giddens gives considerable weight to unintentional actions within structuration theory. For example, he (Giddens 1984, 7) discusses the influence of unconscious motivation and practical consciousness on how people act. As discussed here, the habitus of people's unconscious knowledge of how to live their daily life is an important part of Bourdieu's (1977, 72) understanding of how and why people act. However, with the discussions of environmental and artefact agency in the work of Ingold, Gell, Latour and Hodder, there are different options for describing how things act which do not invoke intentionality directly. I think that the important linking principle here is that it is useful to shift the debate. Instead of asking if a thing can possess agency, which has the danger of reifying 'agency' as a discrete social force, we need to ask how do things act. Latour (2005, 71) does this when he defines the actant as something that makes a difference, as does Gell (1998, 17), with his insight that object agency does not have to be part of a 'philosophically defensible system of thought about agency'. Ingold (2007) makes a similar shift in emphasis when he calls for the abstracted study of 'materiality' to be replaced by a more focussed understanding of the properties of materials. He argues (Ingold 2007, 12–14) that materials do not 'have agency' as an intrinsic property. Instead, they act in the way they do because their physical properties form part of an unfolding environment with the people and things around them. This brings me to another point I wish to develop, which is to propose that we use this concern with how things act to study the actions of mourners, dead bodies and caves in a unified way. All of the elements of thought about agency, when we visualise them in specific examples, are actual tangible things and people. This I take to be the point being made by Fowler (2013, 31–32 and see also Fowler and Harris 2015, 130) about the reality of ideas, objects and the relationships between them. When people act, their bodies do things. They build, dwell and create artefacts. They do this in a material world which they understand and which enables and constrains their actions. The things they make and use persist. In persisting, they form the structure and environment in which other things and people act.

The strongest link of all within all the theory I have considered arises from the common concern with recursive organisation noted here. Unless the passage of time is experienced, Giddens' duality of structure does not happen. Ingold's dwelling perspective does not function without temporality. Gell's object agency is only distributed when objects are observed after they have been created. Latour's kettle has to exist before it can afford the possibility that it can boil water. Hodder (2012, 193), is particularly clear that the material legacies of earlier things, persisting through time, are a central part of 'entanglement'. Thus far, I have presented the actions of time and memory in a relatively uncritical way. Processes take place 'over time', and people and things respond in various ways to the evidences of temporal change. However, as the examples from the ethnographic literature discussed in Chapter 3 show, the way that people understand and relate to the passage of time varies. To properly understand the agency of people and objects, it is necessary to understand how the passage of time is experienced.

Time and memory

There is an extensive literature in both cultural anthropology and the wider social sciences about the human experience of time. There is not the space or the necessity to review all of this work here; however, there are certain recurring themes which it is helpful to explore. When philosophers and anthropologists have written about time, they have almost invariably approached the subject by creating two opposing categories of time or time experience. Confusingly, no two writers use the same terms or even oppose equivalent concepts, but the idea of binary oppositions persists, even if only as a rhetorical device to be overcome. The aim of this section of the book, in line with the recent trends in social thought discussed here, is to try and move beyond these binary modes of thought about time. This ambition has been shared by many earlier writers, but despite this, a remarkable number of binary descriptions of time still occur in the literature.

Gell (1992, 14–36) has provided a review of the array of different binary descriptions of time which have developed in anthropology following from Durkheim's (1995, 9–10) insight that the human experience of time is culturally constructed. These include Evans-Pritchard's (1940, 95–108) contrast between the 'œcological' time of day-to-day and seasonal pastoral activities and the 'structural' time of lineage, descent and age-set succession among the Nuer. Œcological time is process-driven and experiential, whereas structural time is abstract and transcends individual experience – it is mythical, and it does not pass sequentially. Levi-Strauss (e.g. 1963, 301) established a distinction between 'diachronic time', that is to say successive and process-driven historical time, and 'synchronic time', which is cyclical and mythical ritual time. In this case, whole societies were said to organise themselves differently depending on the kind of time which underpinned them. Leach (1961, 125–126) also set up two slightly different categories of time: firstly, alternating reversible events based on repetition of many natural phenomena; and secondly, linear time based on the inevitable and one-directional change and decay that organisms experience over their lives. Secular time was linear, but sacred time was alternating, concerned with reversing the effects of time's flow for ritual and religious ends. Bloch (1977, 284–285) made a similar distinction between what he referred to as 'durational' time, used for practical activities, and cyclical or 'static' time, used in ritual and formal situations.

Comparable binary categories of time can be recognised in discussions within history and archaeology. These often appear to be versions of the contrast between the physical 'time of the world', derived ultimately from Aristotle, and the phenomenological 'time of the soul', originating with St Augustine of Hippo. In his critical review of this tradition, Ricoeur (1988, 21) described this as the distinction between objective and subjective time, neither of which he would regard as an entirely satisfactory description. Within archaeology, Shanks and Tilley (1987, 128) made a similar distinction between abstract and substantial time. Thomas (1996, 34–36) characterises much historical and archaeological thinking about time as 'periodicity'. He gives the example of Fernand Braudel's tripartite chronological systems in history. These are thought of as containers for the human action in the past which is being described. They are examples of objective, scientific time. The effect of this conception of time as a container, according to Thomas, is to focus archaeological and historic analysis preferentially on long-term processes. Harris (2017, 128) has made a similar point about scales of time analysis in both Braudel's *Annales* tradition and the 'time perspectivism' of Geoff Bailey. He argues that both these approaches leave the relationship between the claimed different scales of time under-investigated and have the effect of prioritising the long-duration events as the 'true' scale of analysis. As Thomas (1996, 38) notes, this essentially prioritises scientific objective views of time over subjective, culturally constructed time. The alternative way of treating the passage of time in history, according to Thomas (1996, 38–39), is typified by the networked genealogical histories of Foucault (1979, 152 for example). Harris (2017, 128–130) identifies the problem of different conceptions of time as arising from an unspoken assumption that the processes causing events to happen are fundamentally different. He argues that we need to find ways of thinking about the experience of time which treat the differences between short-term and long-term events as relative differences rather than absolute incompatibilities.

The main lesson from the categories of time reviewed so far is that binary oppositions are unhelpful. For example, as Gell (1992, 29) points out, Levi-Strauss's suggestion that synchronic time is the time of rituals is undermined by the fact that both synchronic and diachronic events can be structured by ritual concerns. He makes a broader point that societies that describe cycles in time do not necessarily have a 'cyclical' view of time overall.

> The relevant distinction does not lie between different 'concepts of time' but different conceptions of the world and its workings … it is equally essential, both to the belief that 'the world goes on and on being the same' and to the contrary belief that 'the world goes on and on becoming different', that one believes that the world goes on and on. (Gell 1992, 36)

From this standpoint, Gell (1992, 149) argues that the problem with much anthropological theory about time is that it has attempted to provide foundational philosophical statements about the 'nature of time' rather than discuss the lived experience of time. He characterises philosophical descriptions of time as being concerned with either physical or phenomenological time (Gell 1992, 150). However, he asserts, from an anthropological point of view, that only descriptions of the human experience of time, phenomenological time, are relevant. Gell's response to the problem of the overly metaphysical nature of much anthropology of time may seem somewhat perverse. He devotes a good part of the centre section of *The Anthropology of Time* to expounding a philosophy of time based on yet another binary opposition between two types of time, the A and B series of time. The key to understanding this apparent contradiction is to note that Gell (1992, 149–150) claims to be looking for a useful rather than a foundational philosophy of time. From this perspective, *The Anthropology of Time* can be seen as a pragmatic project, in the sense that the term is employed by Rorty (1991, 1–6). Like Rorty, Gell (1992, 150–151) has approached the phenomenological literature in terms derived from analytical philosophy – in this case, Mellor's (1981) *Real Time*. This gives him a rhetorical framework to counterbalance what he sees as the unhelpfully grandiose claims of transcendental philosophy; see, for example, his two-page dismissal of almost the whole of *Being and Time* (Gell 1992, 264–266).

This attempt to provide anthropologists with a useful description of the philosophy of time has been extremely influential, particularly in Ingold's (1993, 157–158) definition of temporality discussed here. Gell's (1992, 151–155) binary classification of the A and B series of time is relatively easy to outline, but it has complex consequences. A-series time differentiates between events based on whether they are past, present or future events. B-series time orders events as being either before or after one another. The distinction between A and B series time is explained by Gell (1992, 154–155) in the following terms. The existence of an object in B-series time can be envisaged as a 'linear streak' in space-time. Space-time itself is stable and always present, but we encounter events in a particular order as we move through it, giving the 'before' and 'after' relational qualities which are characteristic of B-series time. In A-series time, by contrast, reality only exists in the present. The present moment is envisaged as an infinitely thin 'screen' in which events exist, but also in which they have evidence of their past states and prefigure their future existence. This distinction can also be thought of in terms of how people describe events in time. In the A-series, events are described in a way which relates them to the present; they are past, present or future events. This means that the truthfulness of this description changes with the passage of time. B-series events are described in terms of their fixed temporal relationship to other events; they occurred before or after another event. Therefore, the truthfulness of the statement, for example the date of an event, does not change as time passes. A-series time is the lived experience of time, of moving continuously from one present to the next, whereas B-series time is the descriptive recording of time to form an intelligible calendar of events.

Both the A-series and the B-series theories of time deal with the human experience of time. However, the importance of the B-series theory of time is that it is a description of this experience. As a description of time, it is therefore social time, an explanation of how one event relates to another, which is aimed at an audience. A metaphor that has been used by other writers (for example Ingold 1993, 157) to summarise the B-series has been to describe events as being spread out like beads along a string. This can give the impression that B-series descriptions must be of linear time. However, as noted here, this does not preclude societies from describing and referring to cycles in time, provided that the events are still described as being either before or after other events. Gell (1992, 89–90) argues that interactions with 'nature beyond society' profoundly influence how people conceptualise time. The habitus and taskscapes of Umeda sago farmers in New Guinea and Muria Gond rice farmers in central India are so fundamentally different as to account for the different way these people express and relate to temporal concepts. He returns to this point in his critique of Bourdieu's analysis of Kabyle conceptions of time. It is possible to read Bourdieu as arguing that traditional people such as the Kabyle have a fundamentally different temporality because of their different habitus. They exist in the flow of A-series time, whereas Westernised people consciously turn their experience of time into a B-series, chronometric account of time (Gell 1992, 290–291). However, Gell (1992, 291–292) goes on to argue that all people create B-series accounts of time. The difference made by their different habitus is a difference in the kind of references they use to describe the relative positions of events in their mental map of time.

This begs the extremely important question: how precisely do we define an event? Gell (1992, 154) quotes Weyl to provide a definition of an event in the B-series as something that is perceived by a conscious agent. This, of course, returns us to the questions considered in the previous section as to how we define agency and agents. I am going to suggest, from the perspectives established in that section, that we regard an event as something that happens to anybody or anything in a network (or *meshwork*, or *assemblage*). Therefore, the B-series of time should be regarded as a material narrative. It describes time happening to objects and people in space. Ricoeur (1988, 21) makes a similar point when he suggests that the way to move past the unhelpful dichotomy between objective and subjective time is to regard temporality as something which is collectively and culturally experienced. Drawing on the work of Heidegger and Ricoeur has led both Thomas (1996, 79–82) and Ingold (1993, 157–159) to a concern with temporality, the time of the participant. Ingold (1993, 157) specifically relates temporality to the way in which A-series time is experienced as a moving present with traces of past events and intimations of future ones. While I would agree with his assertion (Ingold 1993, 159) that temporality is made up of the tasks being carried out, I would not regard that as incompatible with the description of time as events in the B-series. As I hope I have established in the preceding section, bodies, people and things are all active; they are all examples of the kind of participant that can constitute temporality.

What I am attempting to do in this discussion is to find a description of time and temporality which fits with the models of agency discussed in the previous section. In particular, I think it is important to try to find a way of discussing time which allows us not to take the distinction between active subjects and passive objects too seriously. This is particularly important when it comes to discussing the kind of evidence we have for events in the past. As archaeologists, we need to overcome the tendency that I have noted here to treat one set of evidence, stratigraphy and radiocarbon dating, as a B-series container in which the interpretation of more 'social' evidences, such as material culture patterning, can then take place. Rorty suggested in his essay 'Texts and Lumps' that there should not be different descriptions of the workings of the world for the study of nature and culture. 'If a philosophical doctrine is not plausible with respect to the analysis of lumps by chemists, it probably does not apply to the analysis of texts by literary critics either' (Rorty 1985, 2). Drawing on assemblage theory, Harris (2017, 127) makes a similar point: that there is no need to presume that radically different rules are required for different kinds of temporal analysis. In this spirit, I would suggest that we need a description of the passage of time which is plausible for both a radiocarbon sequence and a ritual calendar. The different habitus of the New Guinea sago farmer and the Bayesian statistician will lead them to identify different material traces as being important when creating a calendar, but both of them are engaged in a similar enterprise. They are creating a narrative of ordered events, what Gell (1992, 292) would refer to as a *B-series time-map*.

To return to the discussion of the archaeological process with which I began this chapter, one important point to note is that we do not need to create another set of dualisms between 'our' material narratives about time, which we are creating in the present, and those which existed in the Neolithic. Following Fowler's (2013, 46–48) analysis of the interaction between circulating references and assemblages in the emergence of our understanding of an Early Bronze Age pottery vessel, it is possible to see time-maps emerging from any kind of temporal evidence. For example, Bayliss, Bronk Ramsey and colleagues

(2007, 2) accept Ingold's definition of temporality as an example of A-series time. However, they argue that this A-series is ultimately underpinned by a B-series, by which they appear to mean chronometric time. In their view, to understand past temporality properly, it is necessary to construct the best absolute calendar of past events. While I would agree with them that such a calendar is important, there is obviously a danger here that chronometric data is separated once again from more interpretive views of time and is relegated to the role of periodic container. This would imply that two different indices of time passing need two ontologically distinct categories of time (see Harris 2017, 128 for a further discussion of this point). I do not think that evidence such as Bayesian radiocarbon models represents 'pure' B-series time any more than temporality represents 'pure' A-series time. Temporal evidence such as bodily decay, cave processes, radiocarbon or stratigraphy are the indices from which temporal narratives are constructed. These narratives emerge from our interaction with all of these different indices, but, following Fowler (2013, 40–42), they do not emerge at random. They are an entangled product of the assemblage of pre-existing indices and our approaches to them.

One way of approaching these material narratives can be illustrated by an example I have published elsewhere as part of a discussion of the embodied and performative nature of social memory. At George Rock Shelter, Goldsland, Vale of Glamorgan, different kinds of events can be seen in a variety of different materials, bodies and spaces within the rock shelter (Peterson 2013, 280). At this site (Figure 4.4), we can see that one of the earliest events at the site was the deposition of lithic debitage. The prior existence of these stone tools in the rock shelter formed part of the physical structure for the next event we have evidence for – the burial of seven individuals. These bodies themselves would have provided a clear index of the passage of time, a material narrative of decay which would have been experienced by anyone visiting the site subsequently. Some of these subsequent visitors were part of another event which was once again marked by the deposition of stone tools and waste. Together with these episodic and more strongly defined events at the site, we can also see more temporally dispersed activity; animal bone and pottery sherds from the site were much more evenly distributed through the stratigraphy. In my previous publication (Peterson 2013, 280), I argued that one way to approach this evidence was through embodied and material networks of practices. Therefore, the final part of this chapter will be an expansion of that argument, trying to trace how connections between bodies, caves and objects can be drawn by understanding how they acted over time.

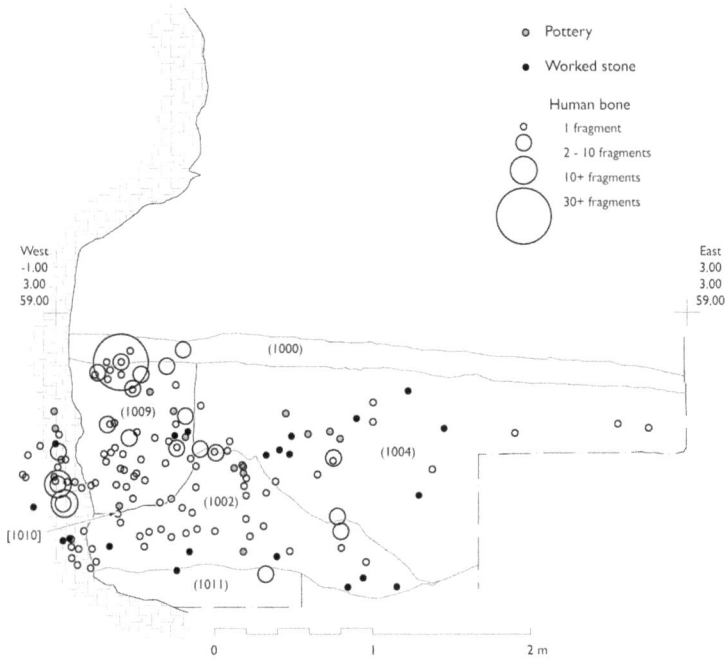

4.4 Deposition of different classes of material culture and human remains at George Rock Shelter, Goldsland Wood, Vale of Glamorgan. Finds from the whole excavated extent of the rock shelter have been projected on a section drawn along the 3 m north line, and therefore some finds appear to lie within the cave wall in this image.

Time and the material world

As time can be regarded as an embodied and material narrative, certain approaches to space and material culture are likely to be helpful in studying how things and people act over time. Any integrated theory of practice needs to bring together living and dead bodies, caves and artefacts. One potentially useful approach is to attempt to broaden the concept of the object biography, which has been the focus of much archaeological research over the last 25 years. The literature around archaeological object biography has been usefully reviewed by Joy (2009), but the theory has its origins in social anthropology (Kopytoff 1986). Kopytoff's (1986, 70–72) original formulation of the idea of an object biography is very strongly tied to the study of exchange processes. His concern is with understanding how different objects are defined by different cultures as belonging in different spheres of exchange. Kopytoff (1986, 72–76) adopts the model that most kinds of exchange can be described as existing on a sliding scale between 'pure' commodity exchange, as exemplified in a monetary transaction, and 'pure' singular exchange, such as the Kula exchange cycle for shell ornaments. His important insight is to see that, just as people understand their role in any society because of what is known of their biography, the appropriate way of exchanging an object is driven by what people know about its previous exchange classifications (Kopytoff 1986, 89–90). In archaeology, arguably the most influential part of Kopytoff's argument was his examples of *how* object biographies can be constructed by comparing the actual use of objects with ideal conceptions of what their use ought to be. His example of the biography of the Suku house is particularly interesting (Kopytoff 1986, 67). The 'ideal' biography of a house as it ages is directly tied to material indices of decay that allow any knowledgeable visitor to infer what the use of any particular house ought to be. Suku houses typically last for about 10 years after first construction, and during this time, they are used in turn as married homes, guesthouses, widow's houses, dens for teenagers, kitchens and, lastly, goat or chicken houses. The physical state of the house, minor decay of the walls and roof for example (Figure 4.5), provides a direct index of the passage of time and hence, of the appropriate use of the house. Kopytoff (1986, 67) states that if a house is 'out of phase' in its use, that is, if its actual biography does not match the ideal biography indexed by its physical state, then a Suku observer would be uncomfortable and

4.5 A material index of the passage of time in an African house wall, in this case an Ilkisongo Maasai house, Kajaido District, southern Kenya.

would draw appropriate and probably unflattering conclusions about the head of the compound. There is an obvious parallel here with the archaeological examples of 'entanglement' in the Catalhoyuk Neolithic houses discussed by Hodder (2012, 192–195), in which the material legacies of earlier constructions and phases of construction create the possibilities for the use of the house at any particular phase.

The use of object biographies in archaeology was developed in a number of different directions in a volume of the journal *World Archaeology* on the subject (Marshall and Gosden 1999). Joy (2009, 541) identifies two different ways in which the papers in this volume, and subsequent archaeological studies, create a methodology for understanding object biographies. The first of these is typically used for the study of portable material culture. In these cases, objects accrue a biography by being exchanged and circulated. For example, Seip (1999) reviews the biography of a single Nuxalk ceremonial mask from the Pacific coast of British Colombia. She is able to show that the mask in question would have originally been part of a highly singular exchange network. When first made, this mask would have formed part of a ceremonial costume which was only revealed at particular events and could only be transferred by inheritance within a lineage (Seip 1999, 277–278). However, the social basis for this singular exchange system collapsed in the later nineteenth century following the extinction of many lineages in an epidemic. This meant that many masks were being held by native people who did not have the rights to either use them or exchange them in the traditional way. It was then possible for masks, such as the one described by Seip (1999, 279–280) to move temporarily into a commodity exchange sphere and to be acquired by ethnographers and collectors. Seip (1999, 280–282) goes on to set out how the mask was then re-singularised initially as an example of ethnographic data and later as cultural property by its new owners, the American Museum of Natural History. The second approach Joy (2009, 541) identifies is more typically used for large static objects, such as monuments and buildings, which are thought to gather biographies by virtue of their persistent presence. For example, Gillings and Pollard (1999, 180–181) examine the way that individual stones within the Late Neolithic henge and stone circle at Avebury act as a fixed locus for the creation of biographies. They focus on a biography of stone 4, which begins with the use of the stone for axe polishing, probably before it was transported to become part of the outer circle at Avebury. Gillings and Pollard (1999, 184–185) describe the movement of the sarsens into Avebury from the wider landscape as a 'gathering' process. From this point on, the stone acts as a material index, not only of the changes which can be seen on its surface, but also for actions which take place around the stone. The stone becomes a fixed material reference point, around which biographical understandings of its immediate setting can be played out. As Gillings and Pollard (1999, 180) note, the stone is not a blank canvas onto which meaning is projected, nor is meaning somehow intrinsic to its material properties. Instead, the material traces in and around the stone index the temporal events from which an object biography can be constructed. From the perspective of attempting to build material narratives, the second of these two approaches is the most likely to be useful when dealing with the prehistoric archaeology of natural places. However, object biographies in this mode can be open to criticism. The open-ended and speculative nature of the narrative makes choosing between different possible accounts very difficult.

Joy (2009, 541–542) makes two significant developments in the way that object biographies are studied in his paper. The first is to recognise the affinity between object biography and other kinds of study of object use-life, such as the *chaîne opératoire* (Dobres 1995, 30–34) and object life histories (Schiffer et al. 2001, 731). While he does not claim that all these approaches are equivalent, he rightly points to the wealth of information which can be added to object biographies from technical studies of object manufacture and use. This focus on the material evidence has the effect of bringing object biography more directly into line with the material narratives I discussed earlier. Ian Hodder (2012, 52–58), makes the important point that one contribution of these approaches is to focus attention on the directionality of many processes. Some things depend on other things to have already happened before they are possible or thinkable. Object biographies in this mode are full of gaps and constraints which provide Hodder with examples of the dependencies within his concept of entanglement.

Joy's second innovation is to move beyond the linear narrative approach of biographies such as Kopytoff's and to think about object biographies as relational, employing this term in a similar sense to Latour (Joy 2009, 545). Just as Latour (2005, 246) cautions against attempts to 'fill in the blanks', Joy (2009, 543–544) recognises that the incomplete nature of the evidence for many prehistoric objects means that a linear narrative will often be difficult to write. Relational and material biographies allow him to treat this as a strength of the approach. Given that a key part of an object biography is the comparison between the 'expected' or 'ideal' biography and the actual practice, then it follows that the necessary detail for both ideal and actual biographies will only be available for some times and places. Joy's (2009, 545) relational biographies require knowledge of the artefact being studied, the wider group of similar artefacts and their wider cultural context, but they do not require knowledge of all of these things for the whole existence of the artefact. One aspect of Joy's (2009, 546–551) study of the Iron Age mirror from Portesham serves to illustrate this point. He is able to demonstrate that the mirror would have been used infrequently to monitor high-status personal appearance and that it would have been kept hidden for much of its life. Nevertheless, both the detail of *how* the mirror was made and its visible form (Figure 4.6) would have indexed a range of different contemporary material culture. As such, it would have linked to material narratives about feasting, combat, food production and the other, rarely seen, mirrors circulating in contemporary society.

A relational object biography, of the kind that Joy develops, may be thought of as having integrated the *chaîne opératoire* as it has been recently deployed in anthropology with Kopytoff's (1986) original emphasis on exchange biographies for objects. Lemonnier (2012, 16) suggests that the study of things in anthropology can be divided

4.6 The Portesham mirror (centre) and the material indices which link it to other Iron Age objects, including shield edge binding; sheet-metal cauldrons; sword scabbards; horse bridles; mirrors and pottery (after Joy 2009, Figure 3).

into two broad traditions. On the one hand, we have 'cultural technology', derived ultimately from the work of Marcel Mauss but particularly associated with Leroi-Gouran's (1994, 230–255) development of the concept of the *chaîne opératoire*. Through the study of the operational sequences involved in making things, cultural technology has attempted to document how the way things are made is connected to the social or ritual context in which they are produced (e.g. Gosselain 1999). Lemonnier (2012, 16–17) contrasts this with a tradition of 'material culture studies' which has focussed much more on the consumption and exchange of objects, within which he specifically includes Kopytoff's (1986) original formulation of object biographies. In a similar way to Joy (2009, 543), Lemonnier (2012, 17–18) argues for cultural technology as an integrated approach to material culture which draws on both these traditions to cover manufacture, use and exchange.

A particular example of this approach returns us to the Ankave eel traps (Figure 4.3) discussed by Gell (1996, 32–34). The original fieldwork on which Gell's argument was based was carried out by Lemonnier (1993). Lemonnier (2012, 45–62) returns to the Ankave eel trap to provide one of his examples of the cultural technology approach and to review Gell's interpretation. Lemonnier (2012, 47–54) documents the complex operational sequences involved in the construction of the Ankave eel trap. As artefacts, they are both much stronger and more technically sophisticated than other eel traps from New Guinea. Lemonnier (2012, 59) argues that we need to avoid the temptation to think that Ankave eel traps are 'ritual' because they are more complicated than we would expect a functional object to be. The traps 'function' to catch eels to be eaten at funerary feasts, a role that it is difficult to ascribe definitively to either a ritual or mundane world. Similarly, the operational sequence of preparing a trap involves actions which we could describe as practical or ritual or as referencing the wider cosmology of the Ankave about eels. According to Lemonnier (2012, 59), the eel trap does not just passively reflect Ankave origin myths and ideas about eels but, because of its physical presence and the embodied processes involved in its construction, it creates them. The trap distributes the agency of the trapper, in the way that Gell (1996) suggested, but its physical presence in the processes of the Ankave funerary ritual provides the structure for the recreation and reiteration of both the ritual and the wider Ankave cosmology.

This discussion may seem to have brought us some distance from the actions of caves and dead bodies, all of which, I hope I have shown, act, but none of which are 'made' in quite the same way as a mirror or an eel trap. However, whether we refer to this style of study as a relational object biography or as the study of cultural technology, it becomes more broadly applicable when we think more critically about what we mean when we talk about making things. Ingold (2000, 339–348) describes the process of making things in a radically different way by treating manufacture as a special case of a wider phenomenon of 'weaving'. Material culture, in this view, is not a static end product, but rather an index of a temporal process. An artisan is engaged with a material which has a particular set of physical properties. They need to bring learnt and embodied skills to bear on that material in a knowledgeable way. They also need to respond to the material results of decisions they have already made; therefore making, in this way, is an extremely good example of the embodied and material narrative discussed here (Ingold 2000, 346–347). This is also a process which extends beyond the final production of any one artefact. Ingold (2000, 347) cites the example of the Yekuana basket makers of Venezuela, who regard all examples of the interaction of people and the manufacture of objects as part of a wider process of weaving the world. Of more direct relevance still is Ingold's (2007, 6–7) characterisation of the permeable boundary between things which are made by people and things which are used by them. He uses the example of cave dwellings in contemporary China to point out that any space, whether it is a built house, a naturally occurring cave or a hybrid of the two, which is dwelt in by people will have the same narratives of material engagement. What is important here is not whether things are made or things act, but that they index the passage of time.

Structure, agency and environment

To return to the question with which I began this chapter, I hope I have shown that we can think in a consistent way about how it is that caves, dead bodies, material culture and living people act. If we think of all these kinds of things as acting, following Latour (2005, 72) in regarding them as equally important parts of any network, then it follows that we can also consider them all as *participants*, in the sense that this term is used by Ingold (1993, 159). Therefore, caves, artefacts and living and dead bodies are equally able to constitute temporality. There should not be one type of time for nature and another for culture. This idea of temporality is important because the material indices of the passage of time are, as in the example of Lemonnier's (2012, 58–60) eel traps discussed here, the place where the recursive nature of structure and agency exists. The places, objects and bodies which interact within any network are not the substrate on which meaning is built, nor are they passive symbols manipulated in line with mental templates. They are the embodied and material narratives within which caves, bodies, people and things are constituted. I hope to show in the following chapters how relational and entangled biographies of objects, caves and dead and living bodies can assist us in understanding cave burial practices.

5

Origins

Introduction

In this chapter, I want to consider the evidence for the origins of cave burial practice in Britain around the start of the Neolithic period. This is not to suggest that there were no intentional burials in the Palaeolithic and Mesolithic in Britain. However, as discussed in Chapter 1, the data gathered by Chamberlain (1996, Figure 1) and updated by Schulting (2007, Figure 2) shows a significant increase in burial activity which broadly coincides with the beginning of the Neolithic in Britain. In the rest of this book, I will be discussing this phenomenon of 'Neolithic cave burial' using some of the strategies I identified in Chapters 3 and 4. The first question I want to consider is how and when this burial practice began. The wider European evidence which I reviewed in chapter 2 provides two important clues. One is that the transition to farming in some limestone regions, especially in Italy, southern France and Spain, seems to be accompanied by an increase in cave burials. Therefore, it could be argued that Neolithic cave burial in Britain was being adopted as part of the shift to a Neolithic way of life and, further, that people who were becoming Neolithic were doing so because of connections with Spain and southern France. Alternatively, the important fact about cave burial in Europe may be the increase in the practice across the continent in the centuries around 4000 BC. In this case, the increase in cave burial may not have been directly connected with the beginning of the Neolithic at all. Rather, it may have reflected connections between Late Mesolithic or Early Neolithic groups in Britain and Middle Neolithic people in Belgium and northern France.

The relatively precise chronology which is now available for the first part of the British Neolithic (Griffiths 2011, 2014a, b; Whittle et al. 2011) makes it much more likely that, where we have well-dated sites, we can distinguish between Late Mesolithic and Early Neolithic origins for burial practices. However, it is not sufficient to simply provide an estimate for the start of burial activity and compare that estimate with the one for the beginning of the local Neolithic. The transition from hunting and gathering is likely to have been a complex and multi-stranded process (Cummings and Harris 2011, 371–375). Some groups of people in any given area may have continued to hunt and gather, while others began to farm. Other groups may have returned to hunting or adopted pastoralism but relied on exchange networks which included arable farmers. Therefore, it is necessary to look at cave burial practice around the fourth millennium BC through the types of relational, material and embodied narratives discussed in Chapter 4. By understanding the detailed history of each site, we can start to understand how different burial practices related to each other and to wider Late Mesolithic and Early Neolithic burial practices. The locations of the sites discussed in this chapter are shown in Figure 5.1.

Late Mesolithic human remains in Britain are rare, but, as Hellewell and Milner (2011, 62) have pointed out, they are not completely absent. Of particular interest here are the human bones which began to be deposited on the latest layers of Mesolithic shell middens such as

5.1 Location map for the sites discussed in Chapter 5. The base mapping includes data licenced from © EuroGeographics.

Cnoc Coig, Oronsay; and An Corran, Skye (Hellewell and Milner 2011, 64). Meiklejohn and colleagues (2011, 36) list one further site where human bone has been dated to the fifth millennium BC: Caisteal nan Gillean II on Oronsay. Rosen (2016, 129), in reviewing the skeletal evidence from cave sites for the whole of the later Mesolithic, tentatively suggests that a successive inhumation rite was practiced at Potter's Cave, Caldey Island; and Foxhole Cave, Paviland, Gower. Hellewell and Milner (2011, 62–63) also suggest that the late fifth or early fourth millennium BC dates from Fox Hole Cave, Derbyshire (Meiklejohn et al. 2011, 38), which have been regarded as Early Neolithic, are equally likely to be Late Mesolithic. Unfortunately, as discussed in Chapter 1, the dates on the Fox Hole Cave, Derbyshire, human remains are among those which should be regarded as unreliable owing to ultrafiltration contamination (Bronk Ramsey et al. 2004). Schulting and colleagues (2013, 22) argued that Hellewell and Milner (2011) were therefore wrong in their general assertion that fourth millennium BC cave burial began before the Neolithic and specifically described cave burial as an independent Neolithic development. However, there other caves beside Fox Hole Cave with very early fourth-millennium BC dates. Hellewell and Milner's general point is potentially still valid for these caves. Several of the sites discussed in this chapter could, based on radiocarbon evidence alone, belong to either the latest Mesolithic or the earliest Neolithic.

Two kinds of site are likely to be helpful in understanding the adoption of cave burial. There are a very few sites where a clear link to known Mesolithic practices can be demonstrated. Because of the extreme rarity of Late Mesolithic human remains, this evidence is very localised. It is essentially confined to a small region of western Scotland, but in that region, it offers the best route to understanding the kind of relational embodied narratives I discussed in Chapter 4. I will consider these sites in detail here to demonstrate how caves, dead bodies and living people acted together to develop one particular kind of Early Neolithic cave burial rite. In the rest of Britain, the relationship of the earliest of these cave burials to Late Mesolithic burial practice is unknown. Therefore, in these cases, I have begun by identifying sites with multiple radiocarbon dates where the earliest burial appears to predate the likely date for the first Neolithic activity in that region. As Hellewell and Milner (2011, 62–63) point out, some of these sites may be genuine Mesolithic precursors to the wider Neolithic practice of cave burial. Alternatively, they may represent an unsuspected early manifestation of a Neolithic way of life. Griffiths (2011, 80–81), for example, excludes several Derbyshire cave burial sites from her regional chronology for the start of the Neolithic on the grounds that, without the presence of diagnostic material culture, we cannot know whether these sites were used by farmers or hunter-gatherers. I have used the archaeology of these sites to construct material histories relating these early cave burials to other local evidence in an attempt to resolve this problem.

Middens and human remains in the Mesolithic and Early Neolithic

The Neolithic cave burials known from the West Highlands and Inner Hebrides are the best example of a clear relationship between Late Mesolithic and Early Neolithic practices, in that they are all associated with shell midden deposits. The shell middens of Atlantic Scotland are an important group of sites which have long been recognised (Armit and Finlayson 1992; Pollard 1990) as having the potential to inform us about the regional transition between the Late Mesolithic and Early Neolithic. Some of these middens are found within caves and rock shelters, and some of them contain human bone.

A recent review of these sites (Milner and Craig 2009) has established a number of important questions about the relationship between human bone, marine shell middens and caves. Milner and Craig (2009, 176–77) demonstrate that human bone deposition is invariably a late phenomenon on these sites, post-dating 4000 BC in all cases. This contrasts with the dates of the middens themselves, which generally begin much earlier. This has led to the suggestion that burials postdate the use of the middens and indeed to a distinction being drawn between Mesolithic middens and Neolithic burial at the same site. Armit and Finlayson (1992, 669) suggest that it was the association of human remains with caves that was important, and that by the Neolithic, the shell middens were of little significance. This apparent separation between Mesolithic middens and Neolithic burials has since been strengthened by stable isotope results from multiple sites, which show a decisive shift away from marine foods at the end of the Mesolithic (see Richards and Schulting 2006 for a review).

The midden at Cnoc Coig, Oronsay (Mellars 1987), is important in this discussion. It is a well-excavated and dated site where human bone is associated with an open-air midden. It has also been the subject of a detailed osteological analysis specifically focussed on attempting to reconstruct the burial process (Meiklejohn et al. 2005). The site was excavated between 1973 and 1979; 196 m^2 of the midden deposit was excavated, which is estimated to have been around 70% of the original volume (Mellars 1987, Figure 14.2). The midden seems to have three phases, the first two of which formed around a large, repeatedly used, central hearth and 'hut-like' structure. The third phase was less structured and may at least partly have been the result of the disturbance and re-arrangement of existing material (Meiklejohn et al. 2005, 87–88). There are forty-four individual fragments of human bone from the site, all from adults apart from one juvenile vertebra fragment. Detailed spatial analysis of the bone distribution has shown that there were at least two separate depositional practices at Cnoc Coig. Dispersed fragments including large amounts of teeth are similar to assemblages from many Northern European Late Mesolithic middens (Meiklejohn et al. 2005, 89). However, the bulk of the bone is grouped into discrete 'bone groups'. The two largest of these, group 2 and group 3, can be shown to represent a different burial practice defined by the presence of large amounts of hand and foot bones. Meiklejohn and colleagues (2005, 96) also identify this burial practice at the much less extensively excavated Oronsay midden sites of Caisteal nan Gillean and Priory Midden.

Milner and Craig (2009, 177) suggest that the Cnoc Coig human bone should be dated to between 4200 and 3650 BC. This date range spans the likely date of the transition from the Mesolithic to the Neolithic in western Scotland. However, Cnoc Coig is particularly important because at this site, burial can be shown to be directly connected with the formation of the midden, that is to say, with a typically Mesolithic practice. Bone group 2 was stratified in phase 1 midden deposits and, although bone group 3 was from phase 3 contexts, part of it was sealed by a hearth, indicating that here too midden activity was still going on after the deposition of this bone (Meiklejohn et al. 2005, 97–98). Most of the human bone has a dietary isotope signature which shows a 'Mesolithic' marine-dominated diet (Richards and Mellars 1998; Richards and Sheridan 2000), demonstrating a connection between the people being buried and the food being consumed at the site. Whatever the precise date of the burials at Cnoc Coig, it appears that the people being buried were hunters, gatherers and fishers and therefore best considered as being culturally Mesolithic.

The Oronsay middens are also important because they have provided clear evidence of use at particular seasons and times of the year. Based on a study of sagittal otoliths in Saithe, the dominant fish species in the bone assemblages, Mellars and Wilkinson (1980, 33–6) suggest that closely defined seasonal occupations can be identified at three Oronsay middens and that these persisted throughout the occupation of the sites. Cnoc Sligeach was probably in use from June to July, Cnoc Coig from September to November and Priory Midden in mid-winter. When other faunal evidence is included (Richards and Mellars 1998, 180–181), then Caisteal nan Gillean can also be seen to have been occupied in early summer. If this pattern is repeated at other middens with burial – and possible seasonality has been identified at An Corran, Skye (Saville et al. 2012, 59) – then we should probably see midden burial as tied to specific seasonal events at each different locality.

Neolithic cave burial associated with midden deposits, therefore, may have had its roots in a Late Mesolithic practice. At the very least, we can see it as being structured by the physical remains of Mesolithic shell-midden accumulation. There are four questions about these burials which need to be considered in more detail. Firstly, is the appearance of burial in middens directly linked to the transition from the Mesolithic to Neolithic? Secondly, is the rite primarily associated with middens or with caves? Thirdly, what is the significance of the shift from depositing faunal remains to depositing human bone? Finally, is there evidence for this rite over a wider area than in western Scotland? There are four cave sites in Highland Scotland with radiocarbon dates on human bone which ought to fall within the Neolithic period. One of these, Reindeer Cave, Inchnadamph (Appendix 1, number 41), in the central mountains of Sutherland, does not contain shell midden deposits. It is Middle Neolithic in date and is therefore discussed in more detail in Chapter 7. The three dated Neolithic shell midden caves are An Corran (Appendix 1, number 1), on the north coast of Skye; and two sites close together on the coast of Argyll, Carding Mill Bay 1 (Appendix 1, number 11) and Raschoille (Appendix 1, number 40).

Carding Mill Bay 1

Carding Mill Bay 1 is a rock shelter which contains a shell midden. It is on the mainland coast of the Sound of Kerrera, about 1 km southwest of Oban (NGR NM 4874 2935). It was excavated under salvage conditions after it was discovered during construction work. The rock shelter is formed of a small fissure in a conglomerate former sea-cliff. Within the fissure were the truncated remains of a shell midden which had been overlain by later prehistoric deposits. Radiocarbon dates on worked antler and bone (Hedges et al. 1993, 311) and charcoal and shell (Connock et al. 1993, 30) from within the midden suggest that it was largely formed during the Neolithic, although some of the dates from the earliest contexts may be Late Mesolithic.

The disturbed remains of a sandstone cist in the upper fill of the fissure contained human bone and a sherd from a food vessel. This is likely to be the remains of an Early Bronze Age burial of one adult woman and one child, although none of this bone has been radiocarbon-dated (Connock et al. 1993, 29). There were further pieces of human bone from beneath this cist (see Figure 5.2). Excluding the woman and child from the cist, there is a minimum of three further individuals from the site: two adults and another child. Bone from both the fissure deposits and the lower contexts of the shell midden has been radiocarbon-dated to the Neolithic (see Appendix 1).

The human bone from the earlier part of the midden dates to slightly later than the worked antler, charcoal and marine shell from the same deposits, which calibrate to between 4200 and 3600 BC (Bonsall et al. 2012, Table 2.1). Given these dates, it is likely that the Neolithic bodies were placed in the top of a midden which had already been in existence for some time. Both the recorded stratigraphy and the radiocarbon evidence suggests that the burials here took place intermittently over two phases. Modelling these sets of data together in OxCal4.3 (Bronk Ramsey 2009; Reimer et al. 2013) shows that the burial in the early shell midden deposits took place over a period

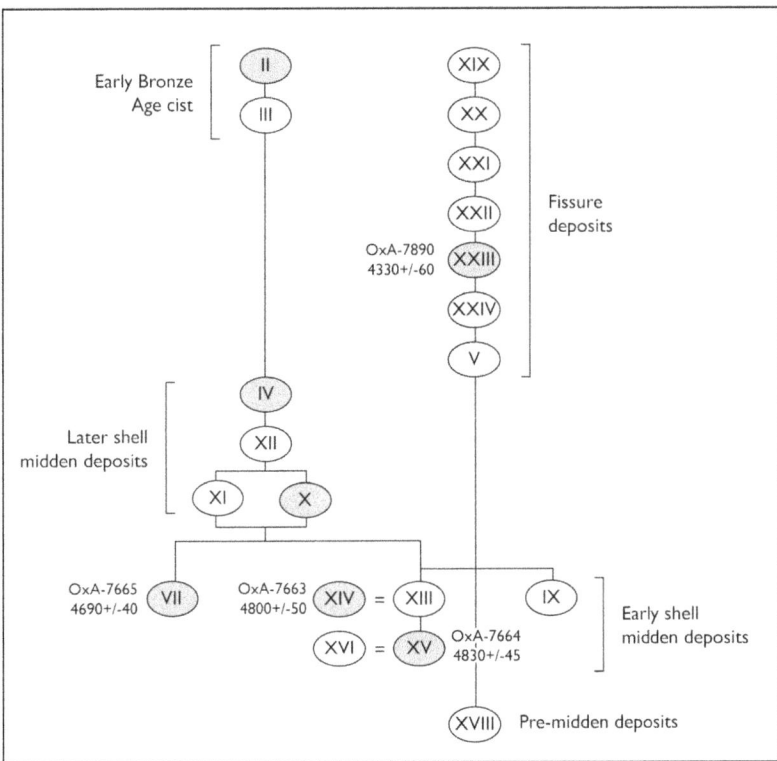

5.2 Relationships between the excavated contexts and dated human bone at Carding Mill Bay 1. Human bone was present in all the contexts shown in grey on this diagram. Based on information in Connock et al. 1993, 27–9 and Bronk Ramsey et al. 2000, 461.

of up to 215 years, starting between 3690 and 3525 BC and ending between 3630 and 3380 BC. After this activity, there was an apparent hiatus before the burials in the fissure which took place between 3330 and 2875 BC. Therefore, we can see both Early and Middle to Late Neolithic funerary use of the Carding Mill Bay 1 cave.

Meiklejohn and colleagues (2005, 101–102) compared the make-up of the bone assemblage at Carding Mill Bay 1 with Cnoc Coig and concluded that it was extremely likely that the same rite was taking place at both sites. This is particularly interesting as, unlike the Cnoc Coig bone, stable isotope results on the Carding Mill Bay 1 material show that these individuals had an almost completely terrestrial diet (Schulting and Richards 2002b, 155–157). The presence of large numbers of hand and foot bones shows that the Early Neolithic rite was some form of successive inhumation on the surface of the midden. If this was an exactly similar rite to the one at Cnoc Coig analysed by Meiklejohn and colleagues (2011, 101–102), it demonstrates that the taphonomic processes which tend to lead to the early disarticulation of hands and feet were being deliberately drawn upon. Palaeotaphonomic research tends to group all the bones within, for example, the hand as equally liable to disarticulate early (Knüsel 2014, 32). However, neotaphonomic studies provide many examples of whole hands and feet becoming disarticulated as a unit (see Dirks et al. 2015, 26–30 for a review). This implies that living people were engaging with the decomposing bodies, drawing on the decay process but also managing it to ensure that hands and feet could be placed appropriately.

The discovery of the cist demonstrates that there were probably further burials at the site in the Early Bronze Age. The idea of this midden as a burial site persisted in some form, although the details of the burial practice were different. I have discussed elsewhere how the visible presence of human bone would have ensured that some memory of the mortuary associations of a site were remembered (Peterson 2013, 276–277). Other caves and rock shelters in the region were also used for burial in the Early Bronze Age: at Benderloch rock shelter, Argyll and Bute, salvage excavations recently discovered evidence for an early second-millennium BC burial associated with a bowl-shaped food vessel from within a midden deposit (Dunbar and Thoms 2008, 10).

Raschoille

This site is another which was recorded under salvage conditions, in this case by the Lorn Archaeological and Historical Society in

1984 (Connock 1985). Raschoille Cave was uncovered from behind a talus slope of angular rock debris during building works on the north-west side of Glenn Sheileach (NGR NM 8547 2888). It has been stated (Milner and Craig 2009, 170) that the midden deposits at this site were very insubstantial, based on the recorded mass of the shell recovered, which was only around 11 kg. However, Connock (1985, 7) states that the aims of the salvage excavation were solely the investigation and removal of the human bone. Excavation ended at the base of the layers containing the bone with two small exploratory trenches into the top of the layer below to confirm that no further human bone was present. So it seems likely that there were substantial midden deposits at this site, even if they were not excavated and studied in detail.

All the dated human remains were recovered from within this rock debris layer (Bonsall et al. 2012, 18) at the entrance to the small fissure cave. There are fourteen dates in all (Bonsall 2000, 112 and see Appendix 1). Three of these are described as coming from lower scree deposits, with the remainder coming from the upper deposits. Detailed osteological data about these remains has not yet been published, but the burials included both adults and children. The three dates from the lower deposits are actually among the most recent from the site. It therefore seems likely that all the dates on human bone should be treated as part of a single sequence. The dates have been modelled in OxCal4.3 (Bronk Ramsey 2009; Reimer et al. 2013) on this assumption. This suggests that the first burial on the site took place between *3795* and *3675 BC* and the last between *3495* and *3200 BC*, with an overall span of use for the site of between 195 and 530 years.

There are no artefacts from the midden which are likely to date to the same period as the known burial activity. Radiocarbon dates on marine shells, hazelnut shells and worked animal bone from the lower deposits range from the mid-sixth to the mid-third millennium BC (Bonsall et al. 2012, Table 2.1). This shows that parts of the midden are much older than the bulk of the human burials but also that there was some overlap between the creation of the midden and the burials. Connock (1985, 5) reports marine shell throughout the layer which contained the human bone. Therefore, as at the nearby Carding Mill Bay 1, it seems as if human bodies were being placed on the top of an established midden. In view of the fact that only the upper levels of the cave were excavated, it is likely that this midden could have been larger and older than currently suspected. The single arrowhead from the site is an Early Bronze Age barbed and tanged form (Connock 1985, 3), much younger than the dated burial activity, but another indicator of a longer range of activity at the site than that suggested by the burials.

Despite the salvage conditions, a record of the position of all the finds exists (see Figure 5.3) which allows some details of the midden and burial structure to be reconstructed. The fact that both cranial and post-cranial elements are evenly distributed over the whole of the excavated surface of the midden suggests that the original burial rite here was the successive inhumation of complete bodies, although final publication of the osteological data will be necessary to confirm this. The presence of large quantities of cranial bone might indicate that the rite here was different to the one that created the hand and foot-dominated assemblages at Cnoc Coig and Carding Mill Bay 1. It is difficult to be too specific, given the limited reporting available so far for this site, but I would suggest that the burials at Raschoille were placed on the surface of the midden. Unlike at Carding Mill Bay 1 and Cnoc Coig, once this had happened, people were not involved in the process of disarticulation. The journey of the corpse within the intermediary period would then have been left entirely to the agency of the cave and bodily decomposition processes.

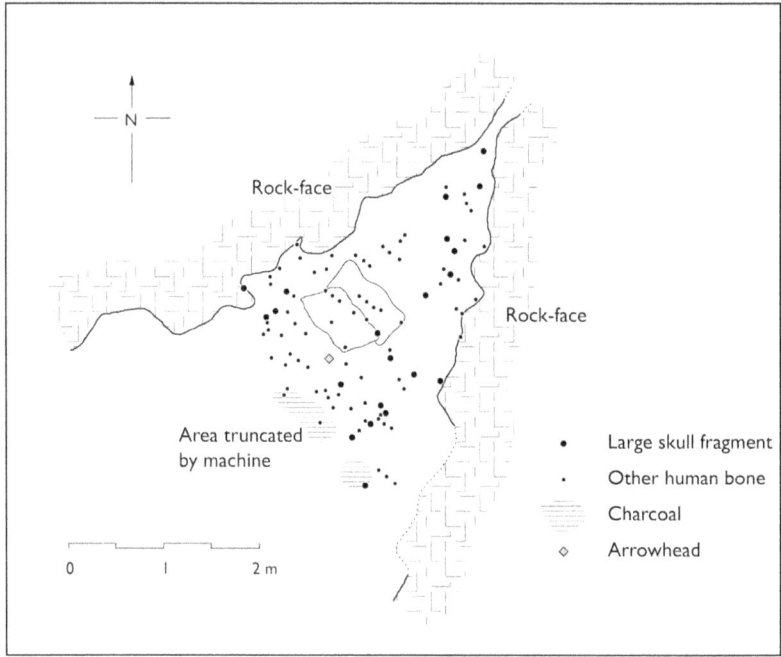

5.3 Plan of the excavated deposits at Raschoille Cave (after Connock 1985, Figures 2 and 3).

An Corran

The rock shelter at An Corran on the Isle of Skye was excavated in the winter of 1993–1994 in advance of cliff stabilisation works (Saville et al. 2012, 3–7). The site is in a sea cliff on the north-east coast of Skye at the east end of Staffin Bay (NGR NG 4915 6848). The deposits in the shelter could be broadly divided into two. The upper layers were largely windblown sands, with a series of hearths providing evidence of occasional human use of the shelter. Beneath this, and only exposed in a relatively narrow sondage, were a complex series of shell-rich midden deposits (contexts 31–41, Saville et al. 2012, 13). These midden layers contained bone and antler bevel-ended tools, worked stone and faunal remains alongside the shell. The radiocarbon evidence indicates that this is another midden which developed over a long period of probably intermittent use from the Mesolithic to the later Neolithic (Saville et al. 2012, 80–81).

Human bone came from two contexts in the upper part of the midden (see Figure 5.4). There were twelve fragmentary pieces of bone and three teeth from context 31 and twenty-seven bone fragments and four teeth from context 36. There are Neolithic radiocarbon dates on five bones from adult individuals (see Appendix 1), three from context 31 and two from context 36. The radiocarbon results suggest that there were at least six burials at the site. The published bone report gives a minimum of five individuals: two adults, one of whom was over 40 years old; a sub-adult; and two children, one aged around 9–12 months and the other around 5 years old (Bruce in Saville et al. 2012, 44–45). This assumes that the cervical vertebra dated as OxA-13552 and the ulna dated as AA-27743 belong to the same individual. However, these two dates do not overlap even at two standard deviations, and therefore it is likely that there were two different mature adult individuals and a total of six burials at the site.

It is highly unlikely that these dates represent a single episode of deposition; all the dates are significantly different. Even if the earliest and latest dates are excluded, the remaining three dates still fail a X^2 test at 5% when an attempt is made to combine them. Modelling the dates in OxCal 4.3 (Bronk Ramsey 2009; Reimer et al. 2013) on the basis that the midden was used episodically for burial events over the full range of the dates suggests that the site was in use for between 915 and 1165 years. The first burial probably took place between *3500 and 3360 BC* and the last between *2475* and *2300 BC*.

At An Corran, there is some evidence that the use of the midden overlapped with the deposition of human bone at the site. Two of the bevel-ended bone tools produced Later Neolithic radiocarbon dates, although the bulk of the animal bone and almost certainly all of the stone tools were Mesolithic (Saville et al. 2012, 33, 74). We should probably envisage the burials at An Corran as being placed either into or on top of a well-established midden. Archaeological evidence suggests that burial appears to have been an intermittent and long-lived activity which was taking place within a wider set of related activities, meaning that some tools, animal bones and possibly lithics were

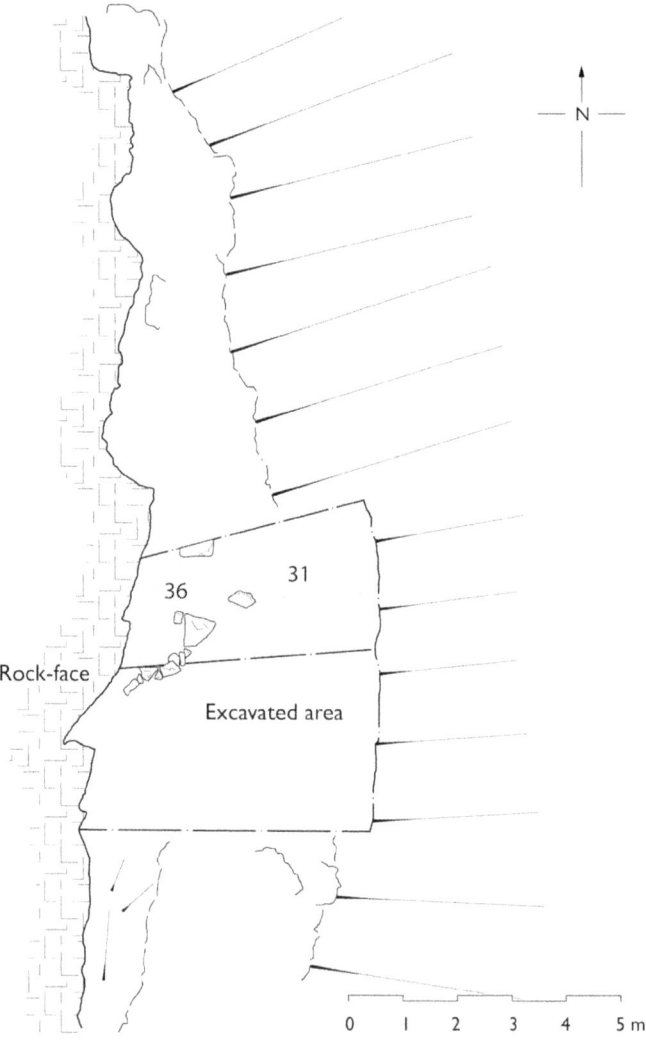

5.4 The excavated area at the An Corran rock shelter, showing the extent of contexts 36 and 31, the upper midden layers where human remains were found (after Saville et al. 2012, illustrations 9 and 15).

still being added to the midden. I have discussed elsewhere (Peterson 2013, 280) how the deposition of different types of material culture could be linked together in biographies of practice. This might arise when objects were being deposited in the same places or as part of the same event, especially if the association was being repeated at particular times of the year or around particular transitions in people's lives. Like other shell middens, An Corran is likely to have been used at particular seasons of the year (Saville et al. 2012, 59). The spatial repetition of putting material on the same midden, linked to calendrical repetition of doing this at specific times of the year, would have allowed the mnemonic associations of the practice to be transferred between food (shell and animal remains), tool use, manufacture (bevel ended tools and lithics) and burial.

Midden burials outside Argyll

Human burial associated with midden deposits is not solely confined to western Scotland, although evidence is extremely sparse in the rest of the country. This is probably a reflection of the different post-glacial environmental history of the north and west. The Late Mesolithic coast in more southerly regions was below current mean sea levels (Lambeck 1995), suggesting that coastal shell midden sites would also now be submerged. Despite this, a few Late Mesolithic and Early Neolithic middens in Britain and Ireland survive which contain human bone. Scattered and disarticulated human remains from a large open-air shell midden at Rockmarshall, Co. Louth (Woodman 2015, Figure 4.28), have been dated to between 4720 and 4360 BC (Hedges et al. 1997, 541). Also of possible relevance here are the partial remains of a woman recovered from Prestatyn, North Wales, in 1924 and subsequently dated to between 3750 and 3535 BC. This woman had a relatively high proportion of marine foods in her diet compared to other stable isotope results for the Neolithic (Schulting and Gonzales 2008). A complex of six shell middens spanning the Mesolithic to Neolithic transition were discovered in advance of housing development at Nant Hall Road, Prestatyn (Thomas and Britnell 2008, 268). These shell middens lie along the edge of the same peat deposit in which the human remains were found. Schulting and Gonzales' (2008, 303) description of the find spot for the human bone would place it at NGR SJ 06569 82932, around 200 m from the edge of the housing development and potentially closer still to other unexcavated middens in the complex. A similar example of Neolithic human remains recovered close to shell middens comes from Sumburgh, Shetland. Here a multiple burial within a cist was found approximately 400 m from shell middens at West Voe; both the middens and burials date to the second half of the fourth millennium BC (Walsh et al. 2011, 3–5).

The only apparent example of a midden burial from a Neolithic cave outside western Scotland is not a shell midden at all. Broken Cavern (Appendix 1, number 9: Roberts 1996), is one of a group of caves in the Torbryan Valley, South Devon (NGR SX 8150 6748). At this site, one stratum of the rock-shelter floor was entirely covered with what was apparently Early Neolithic midden material (Berridge in Roberts 1996, 203). The human remains from this site are a small collection of fragments, dominated by teeth. One of these teeth was radiocarbon-dated (OxA-3206 4885 +/- 90 BP), giving a calibrated date range of between 3942 and 3382 BC. There was an extensive collection of faunal material, dominated by wild species but with a domestic component. Two of these bones have radiocarbon dates (OxA-3205 on a sheep molar 4930+/-90 BP; OxA-3207 on a juvenile cattle tooth 5015 +/-80 BP), which have been modelled on the assumption that they represent a single phase of midden accumulation (Table 5.1 and Figure 5.5). These dates suggest that the midden formed early in the fourth millennium BC. Two very similar dates (OxA- 4493: 5060 +/- 70 BP and OxA-4495: 5010 +/- 70 BP) have been obtained from aurochs teeth from the nearby site of Three Holes Cave (Berridge in Roberts 1996, 203). The Broken Cavern material also included substantial quantities of Early Neolithic pottery: around 200 sherds from five different vessels. There were also two complete stone axes, a leaf-shaped arrowhead and debitage indicating *in situ* working (Berridge in Roberts 1996, 203). The burnt material within this layer was re-deposited (Collcutt in Roberts 1996, 203), possibly indicating that most of the midden material had been moved into the cave from elsewhere. The other interesting aspect of these midden sites is their early date. They are among the earliest dates contributing to the modelled date of *3940–3735 BC* for the start of Neolithic activity for south-west England published by Whittle and colleagues (2011,

Table 5.1 Radiocarbon determinations from cave midden deposits in the Torbryan valley (Hedges et al. 1996, 397–398).

Lab. Number	Element	ID number	$\delta 13$ C (‰)	Date BP	Error (years)	Start (BC 2Σ)	End (BC 2Σ)
OxA-3205	sheep molar	BRKFA 602	−21.8	4930	90	3960	3525
OxA-3207	juv. cattle molar	BRKFA 665	−21.0	5015	80	3960	3660
OxA-4493	aurochs tooth	THRFA 1088	−22.1	5060	70	3985	3695
OxA-4495	aurochs tooth	THRFA 1186	−21.3	5010	70	3955	3660

Origins

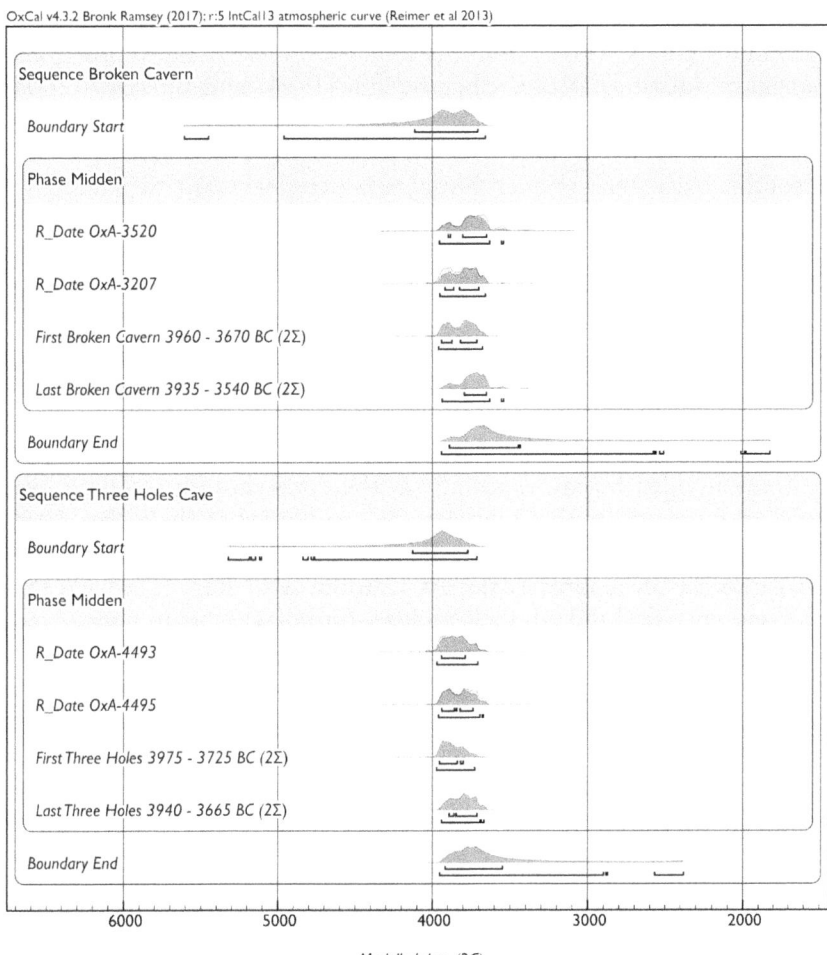

5.5 Modelled dates for midden deposition in caves in the Torbryan valley, Devon, using OxCal v.4.3 (Bronk Ramsey 2009; Reimer et al. 2013).

516–517). An association between terrestrial middens and caves might also be implied by the large assemblage of animal bone from Heaning Wood Bone Cave, Cumbria (Smith 2012b, 6). At this site, cut-marked cattle and pig bones were radiocarbon-dated to the Early Neolithic.

Understanding the cave and midden burial tradition

It is now clear that Neolithic human remains associated with midden material in caves need to be related to a wider debate about the appearance of human remains at midden sites around the Late Mesolithic to

Early Neolithic transition. The evidence from Cnoc Coig, Prestatyn and Sumburgh also shows that it is somewhat misleading to talk about a single midden burial rite. Chamberlain (1996) and Blockley (2005) have established that human bone is extremely rare from all contexts in the period before around 4800 BC. As we have seen here, if it were Mesolithic, then shell-midden burial was extremely late, depending on the exact marine reservoir correction to be applied to the Cnoc Coig material; it may all post-date the start of the fourth millennium BC (Milner and Craig 2009, 175–177). It is therefore extremely likely that the adoption of this new practice or set of practices was one of the suite of changes which we now see as marking the transition between the Late Mesolithic and the Early Neolithic. The question is therefore whether it is possibly to identify 'Mesolithic' or 'Neolithic' cultural practices in any of these burials.

Some midden burial, especially at the open sites on Oronsay, was practiced by people with a high marine component to their diet. These people were apparently fully 'Mesolithic' in the sense of being fishers and gatherers on a large scale. At Carding Mill Bay 1, we see that the same specific kind of shell-midden burial was being carried out by a population with an almost completely terrestrial diet, apparently fully 'Neolithic' and having adopted a farming lifestyle. Two conclusions can be drawn here. First, the continuity of rite strongly suggests that there is some continuity of population in the region between Late Mesolithic Cnoc Coig and Early Neolithic Carding Mill Bay 1. Second, the adoption of Neolithic technologies and social practices in Argyll was an example of the *bricolage* model suggested by Thomas (2003), whereby different elements of a new cultural repertoire were adopted at different times and to different degrees.

Midden burial is not just a cave phenomenon, although Armit and Finlayson (1992, 669) are correct in emphasising that burial middens were predominantly in caves. What we can see is that the sites which have later burial activity are all caves, and that we therefore may have a pattern of a more broadly applicable rite becoming transformed into a specific cave burial rite later in the Neolithic. It may also be that cave middens are over-represented through differential preservation and that, even in areas like Argyll, smaller open-air middens have been lost through erosion. Summarising the radiocarbon evidence from the sites with multiple dates, it is likely that there was burial at Cnoc Coig sometime around 4000 BC. Both Raschoille and Carding Mill Bay 1 have Early Neolithic dates, around 3795–3200 BC and 3690–3380 BC, respectively. After an apparent hiatus in burial activity, there was also a Middle and Late Neolithic phase at Carding Mill Bay 1 (around 3330–2875 BC). The burial activity at An Corran began at a similar date but continued into the very Late Neolithic or Early Bronze Age, approximately 3500–2300 BC.

Midden burial took place within an old context. In almost all these cases, the middens can be shown to have been old structures by the time the first burials took place. The middens themselves were transformed spaces, evidence of long-term human activity at these sites and, particularly, evidence of the consumption of food. They would also have been physical indices of seasonality. Midden burial would therefore have been a new aspect of the use of the sites, but one which drew upon all of these previous meanings. In this case, as the new burial rites associated with the Neolithic developed, they acted to commemorate a set of seasonal activities which were almost certainly no longer practiced. If this was the case, and large open-air middens in particular were no longer forming at the same rate, then the focus of the rite seems to have shifted from middens to middens specifically associated with cave and rock-shelter spaces. However, it is interesting to note that shell-collecting tools were still being added to the An Corran cave midden until the end of the Neolithic.

Other early cave burial evidence

The connection between cave burial and midden burial in western Scotland is the best evidence for a relational link between Mesolithic activity and the development of cave burial. However, as the dates from the excavated cave sites are not particularly early, this does not tell us whether people used caves for burial before the beginning of the Neolithic. In this case at least, cave burial was a Neolithic practice. I now want to follow the second approach to the problem I outlined at the start of the chapter. Rather than look at the date of practices for which we can see a clear relational link to the Late Mesolithic, I will look at the kind of rites which were used on sites where some of the human bone may to be too early to be Neolithic. Following the argument in Hellewell and Milner (2011, 62–63) cited here, this would potentially also increase our knowledge of Mesolithic burial practice. I am aware of six sites with dates which could span the local transition from the Mesolithic to the Neolithic: Bob's Cave, Devon; George Rock Shelter, Goldsland Wood, Vale of Glamorgan; Kinsey Cave, North Yorkshire; Sewell's Cave, North Yorkshire; Spurge Hole, Gower; and Thaw Head Cave, North Yorkshire. However, some of these sites are extremely unlikely to represent genuine Mesolithic burials. The Spurge Hole burial, for example, is dated by a single radiocarbon date (OxA-3815) with a 100-year standard deviation (Schulting and Richards 2002a, Table 3), only a small part of which falls before 3765–3655 BC, the modelled date for the start of the Neolithic in south-west Wales (Whittle et al. 2011, 548). The Spurge Hole burial is also likely to be unsuitable as a dating sample for other reasons which I will discuss in detail in Chapter 6. Figure 5.6 provides a broad comparison between the dated human remains from these caves and the current best model for the local beginning of the Neolithic.

Figure 5.6 was generated by Seren Griffiths, who has added the following comments. 'In the south west, it is 65% probable that the calibrated radiocarbon date OxA-4983 predates the estimate from Whittle et al. 2011 for the start of the Neolithic (*Start south west neo*). In South Wales, it is 82% probable that OxA-X-2424–44 occurred before the regional start of the Neolithic (*StartSouthWales neolithic*) as calculated in Griffiths 2011. In Yorkshire and Humberside, it is

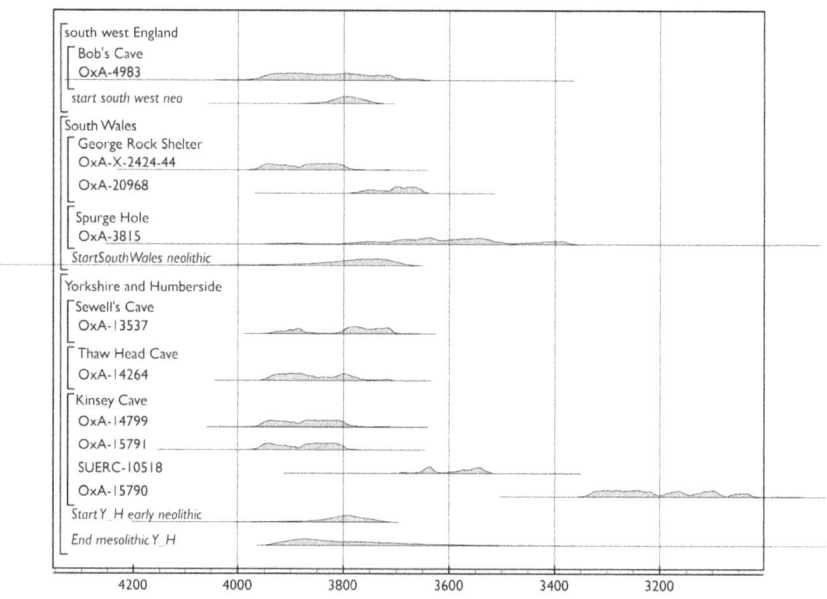

5.6 Radiocarbon-dated human remains from early cave burial sites compared with the modelled posterior density estimates for the start of the Neolithic in south-west England, South Wales and Yorkshire and Humberside, as follows: End mesolithic Y_H (OxCal start boundary parameter estimate for the end of Mesolithic activity in Yorkshire and Humberside) from Griffiths 2014b; Start Y_H early neolithic (OxCal start boundary parameter estimate for the beginning of Neolithic activity in Yorkshire and Humberside) from Griffiths 2014a; StartSouthWales neolithic (OxCal start boundary parameter estimate for the start of Neolithic activity in south Wales) from Griffiths 2011; and Start south west neo (OxCal start boundary parameter estimate for the start of Neolithic activity in south-west England) from Whittle et al. 2011.

75%, 83% and 84% probable that OxA-14264, OxA-14799 and OxA-15791 respectively occurred before the start of the Neolithic for the region as calculated in Griffiths 2014a (*Start Y_H early neolithic*) these radiocarbon dates also most probably occurred before the latest Mesolithic activity from the region as calculated in Griffiths 2014b (*End mesolithic Y_H*).'

At first glance, the most promising candidate for a Late Mesolithic cave burial on this list seems to be George Rock Shelter. However, as discussed here, there are problems with the amount of collagen preservation in one of these dates. At Bob's Cave and the three North Yorkshire sites, the radiocarbon evidence suggests that these sites were used at the time that the transition to the Neolithic was taking place. Whether the people who used these caves were farmers or hunter-gatherers can only be resolved by a detailed examination of the surviving archaeological evidence from each site.

George Rock Shelter, Goldsland Wood, was excavated between 2005 and 2007 as part of a research project investigating Holocene cave use (Appendix 1, number 21: Aldhouse-Green and Peterson 2007). The site is a small, east-facing rock shelter around 6 metres long and just over 2 metres deep. Together with the nearby Wolf Cave, it is one of a pair of sites on a limestone ridge near Wenvoe in the Vale of Glamorgan (NGR ST 1121 7151), both of which have produced Holocene human remains. The earliest excavated deposit in the shelter was an open clast-supported scree, context 1011. This was covered by a thick layer of granular tufa and limestone fragments, context 1002/1007. This in turn was sealed by context 1004, a reddish brown silt with many small and medium limestone fragments (Figure 5.7). The sequence was disturbed by the digging of a large pit close to the rock face at some point during the last 200 years. Finds from the site included a substantial assemblage of animal bone, which has not yet been fully studied but included both wild and domestic species (Ros Coard, personal communication). There were both Neolithic and Mesolithic artefacts from the site (Aldhouse-Green and Peterson 2007; Rosen 2016, 176): fragments of at least one Early Neolithic bowl, a leaf-shaped arrowhead, some Early Neolithic debitage and a small assemblage of three Late Mesolithic microliths. The human bone at George Rock Shelter, except where it had been disturbed by the modern pit, was predominantly found in contexts 1002/1007 and 1004.

There are two early radiocarbon dates on human bone (see Appendix 1) from this site. The earliest of these, OxA-X-2424–44, should be treated with slight caution owing to difficulties in extracting sufficient collagen. However, if we accept this date and model the two dates together in OxCal 4.3 (Bronk Ramsey 2009; Reimer

5.7 George Rock Shelter under excavation in 2007. The very light-coloured tufa-rich layer 1002/1007, which is where the bulk of the human remains were probably originally deposited, is clearly visible in section. Above this is context 1004, which also contained human bone and prehistoric artefacts. Close to the rock wall, the fill of the modern disturbance can be seen as a much darker area in section.

et al. 2013) on the assumption that they represent a phase of burial activity, then we can see that the first burials at George Rock Shelter took place between *3965* and *3780 BC*. In this model, burial activity went on until between *3780* and *3650 BC* (Figure 5.8). This is potentially extremely interesting, as the whole calibrated range for the beginning of burial falls into what is assumed, chronologically, to be the Late Mesolithic period in south-east Wales, and almost the whole of the calibrated range for the final dated burial falls into the Early Neolithic (Whittle et al. 2011, 548). Therefore, George Rock Shelter seems to give us another example of a burial practice which begins in the Mesolithic and continues into the Early Neolithic.

Recent analysis of the human remains from the site (Williams 2008, 46–48) noted that there was a high proportion of hand and foot bones in the assemblage and that all elements were highly fragmented. Williams also established that, although weathering and canid gnawing had taken place, this was relatively limited and suggested that the surviving bone cannot have been on the rock-shelter surface for longer than about 3 years before being buried. On this basis, she suggested

Origins

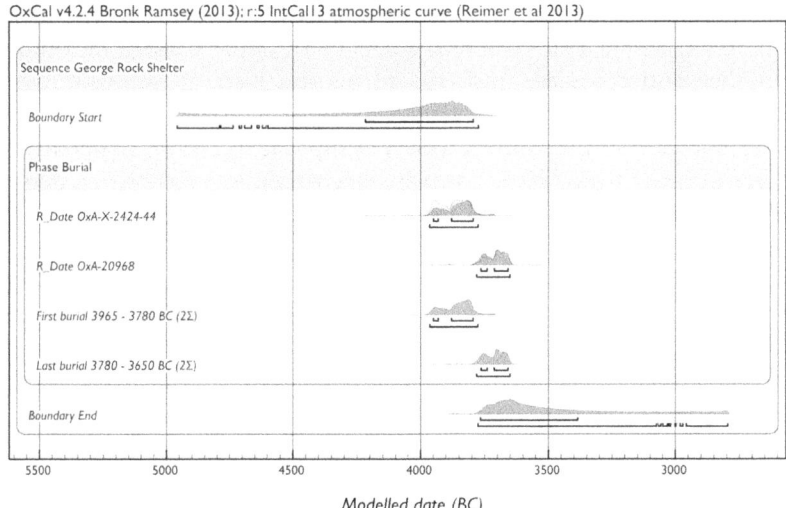

5.8 Modelled dates for the start and end of burial activity at George Rock Shelter using OxCal v.4.2 (Bronk Ramsey 2009; Reimer et al. 2013).

that George Rock Shelter was used to expose bodies during an intermediary period of less than 3 years in a secondary burial rite. The final burial place would have been at another location, and the fragmentary remains left at George Rock Shelter would possibly have been deliberately buried. Williams (2008, 39–40) identified a minimum of eight individuals in the assemblage. This figure was supported by a separate analysis of the dental remains from the site (Tellier 2009). However, since these reports were compiled, radiocarbon dating on one of the teeth has shown that at least one of these individuals was an intrusive post-medieval burial (Higham et al. 2011, 1070). This burial probably took place in the recent pit at the back of the rock shelter visible in Figure 5.7 and would have disturbed existing prehistoric human remains. Because of the extremely fragmentary nature of the bones, it has not yet been possibly to identify definitively which elements belong to the intrusive burial, and therefore the conclusions in this chapter about the prehistoric rite need to be treated with a degree of caution.

The sequence of deposition at George Rock Shelter began with the placing of worked stone at the base of context 1002/1007; this included the Mesolithic artefacts. Burials took place slightly later, within what was probably a rapidly forming tufa and scree layer. If William's (2008, 48) suggestion of a short intermediary period is accepted, then large elements of the disarticulated bodies were removed from this matrix, but the active layer formation would have preserved the remaining fragments. Therefore, George Rock Shelter can be regarded as a site where the agency of living people, decomposing bodies and the scree and tufa formation were all actively drawn on as part of the intermediary period rite.

Human bone from Bob's Cave, Yealmpton, in Devon (Appendix 1, number 6: NGR SX 5739 5124), was discovered in a cave earth deposit inside the cave. The cave has a south-facing entrance and extends for around 20 metres into a limestone cliff. It is part of the Kitley complex in Western Torrs quarry and was partly excavated by John Wright in the late 1980s. The cave was almost entirely filled by a silty cave earth. The single date from this site (Appendix 1) comes from a human femur which was associated with Late Upper Palaeolithic artefacts and animal bones in the cave earth, although no precise find locations have been published (Chamberlain and Ray 1994, 42; Hedges et al. 1998, 437). In view of the limited archaeological and osteological information available for Bob's Cave, the most that can be said about this date is that the cave was used for burial at some point in either the Late Mesolithic or the Early Neolithic.

Kinsey Cave in Giggleswick Scar, North Yorkshire (Appendix 1, number 26: NGR SD 8040 6572) was excavated between 1925 and 1932 by W.K. Mattinson (Jackson and Mattinson 1932, 5). The cave (Figure 5.9) is a relatively large arch in a limestone cliff, leading to a passage which is now around 30 metres long. Mattinson removed a 2-metre-thick deposit of fallen limestone blocks which was masking the entrance to the cave. Beneath this layer, and extending into the cave itself, was a cave earth deposit which contained Pleistocene animal bones, a worked piece of reindeer antler and 'several human bones' (Jackson and Mattinson 1932, 6). The assemblage was reviewed by Lord and colleagues (2007, 687–691) as part of a project examining the Late-glacial cave assemblages from a group of Giggleswick Scar caves. This project obtained a date of 5074 +/- 36 BP (OxA-14799) on a human mandible in the Mattinson archive. This bone was recorded as coming from scree and colluvial material which had probably slipped from the entrance further back into the cave. The date (see Appendix 1) would calibrate at two standard deviations to between 3960 and 3790 BC. Further excavation work was carried out at Kinsey Cave in 2005 (Taylor et al. 2011) and, although this has not yet been published in detail, three more radiocarbon dates have been obtained (Griffiths 2011, 946, and see Appendix 1) on human bone from the same deposit. All four dates have been modelled using OxCal v.4.3 (Bronk Ramsey 2009; Reimer et al. 2013) on the assumption that the burial activity represents a single phase of activity. If this is the case, then the first burial at Kinsey took place sometime between

5.9 The view from the interior of Kinsey Cave across the area of Mattinson's excavations in the entrance and the probable area of Neolithic burial activity.

3965 and *3810 BC*, and burial continued in the cave entrance until *3350–3030 BC* (at two standard deviations).

Unfortunately, detailed osteological information is not yet available for the Kinsey Cave assemblage. Chamberlain (2014) suggests a minimum number of four individuals, with at least two different adults and two different juveniles having been present. The dated samples show that both cranial and post-cranial elements survived in the cave. Therefore, although the radiocarbon evidence suggests, as at George Rock Shelter, that it is highly probable that burial activity began in what is chronologically considered to be the Mesolithic and continued into the Early and Middle Neolithic, we do not have good evidence for the kind of burial rite which was taking place.

Thaw Head Cave, Twistleton Scars, North Yorkshire (Appendix 1, number 46: NGR SD 7105 7590), was discovered during exploration led by John Thorpe in February 1986. The site is slightly to the north of the other known cave burial sites in this part of the Yorkshire Dales. It is a small chamber, 4 × 2 metres in area, accessed through a triangular entrance 0.5 metres high and 0.5 metres wide. The artefacts and human remains from the site were recovered between February 1986 and the end of 1987, as the cave deposits were removed in an attempt to link the cave to the Major Dale Barn cave system. Some recording of the stratigraphy was attempted and the position of finds was noted (Gilks 1995, 1–2 and see Figure 5.10). A layer of large limestone slabs

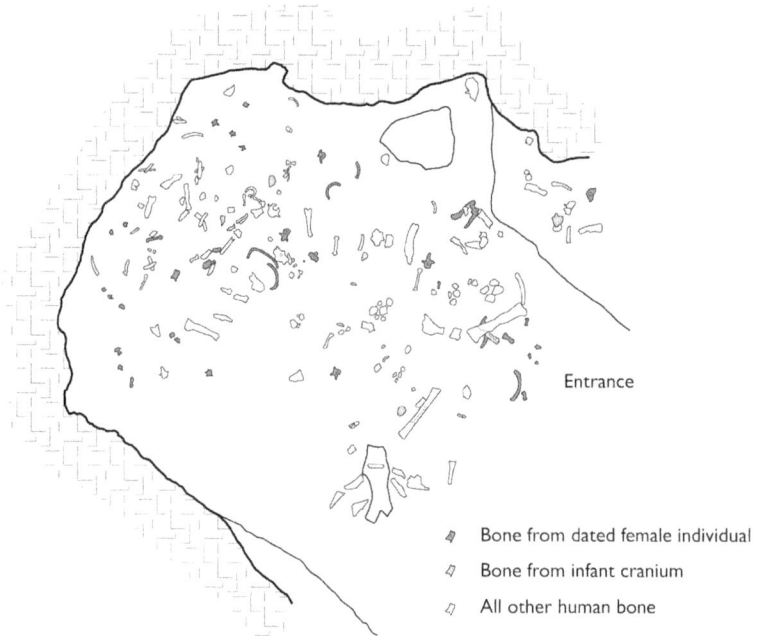

5.10 The recorded position of human bone from within Thaw Head Cave, after Leach (2006, 72).

was discovered in the entrance to the cave. These were interpreted by Gilks (1995, 2), in his reassessment of the archive, as the collapsed remains of a dry-stone wall which had sealed the cave entrance. The human remains were discovered beneath this wall collapse immediately inside the cave entrance and scattered throughout the cave. Some human and animal bone was also found behind another possible dry-stone wall at the back of the cave. Some of the bone was covered with a tufa deposit (Leach 2008, 41). Gilks (1995, 2) suggested that the original site of the burial had been immediately inside the entrance and that, following disturbance, some of the bones had been deliberately redeposited behind walling at the back of the cave.

Thaw Head Cave was one of five sites in the region reassessed by Leach (2006, 2008) as part of her PhD study of the human bone from cave sites. As part of this research, a single radiocarbon date was obtained on the mandible of a young adult female (Appendix 1). This result (OxA-14264: 5040 +/- 31 BP) would calibrate to between 3955 and 3715 BC at two standard deviations.

Leach (2006, 78–83) was able to establish that there were the remains of at least five individuals in the cave, although three of them are represented by single bones. The vast majority of the bones can be ascribed to the dated individual, a woman who was between 17 and 19 years old when she died (Leach 2006, 78). This woman was placed in the cave as a complete fleshed body. There was good representation of the whole body, including the hand and foot bones. There was also very limited evidence for weathering (Leach 2006, 75). Therefore, the body was either buried within the cave sediment or protected by the dry stone walling which sealed the cave. Using the original archive plans (see Figure 5.10), Leach (2006, 75) suggested that this burial took place at the back of the cave chamber. After this initial burial, two different kinds of natural agent acted on the woman's body. The cave floor was a site of active tufa formation at the time of the burial and, as the body decomposed, some of the bones became partially covered in tufa. Leach (2008, 51) has suggested that the particular properties of tufa were known and sought out for burial caves in this region, perhaps in a similar way to the cult of 'abnormal water' identified by Whitehouse (2015, 57–58) in Italian caves. The extensive scattering of the bone, visible in Figure 5.10, was the result of subsequent carnivore scavenging. There are tooth scores, chipping and splintering present on the articular ends of the long bones and the pelvis (Leach 2006, 75–76). Therefore, at some point after the initial burial, her body was no longer protected by the cave sediments or the dry stone walling, and a new phase of bodily decomposition began.

Leach (2006, 75) identified individual 2 as a neo-natal infant, and it is likely that individual 1 died in either while giving birth to this infant or in the late stages of pregnancy. However, the presence of a single bone each from three other bodies within the cave, all of which were also heavily affected by carnivore scavenging (Leach 2006, 75–76), might suggest that the embodied narrative of decomposition undergone by individual 1 at Thaw Head was only the most recent example of this funerary process. Alternatively, it may be that these bones were introduced into Thaw Head cave by the scavenging animals, in which case a similar set of processes may have been taking place in other nearby caves.

Sewell's Cave is a small cave or large rock shelter excavated between 1932 and 1934 by the Pig Yard Club, an extremely active local archaeological group based in Settle. The site itself is in Common Scar, around 800 metres north of the Cave Ha complex (Appendix 1, number 44: NGR SD 7847 6658). At the start of excavations, the rock shelter was almost completely obscured by a thick deposit of limestone blocks which had fallen from cliff above. Once these were removed, the exposed rock-shelter was around 4 metres deep and 3 metres high (Figure 5.11). It extended along the rock face for around 12 metres (Raistrick 1936, 191–192). The Pig Yard Club excavations

5.11 Sewell's Cave, showing the area of the rock shelter against the northern wall, where the human remains were deposited.

defined the stratigraphic sequence in the rock shelter. The uppermost element was a relatively thick layer of clay with limestone blocks. This contained a large number of first and early second-century Romano-British pottery sherds, worked bone and metalwork. There was also some human bone from this layer. Beneath this was a thin layer of 'cave earth', which in turn covered a talus deposit; beneath the talus was a layer of glacial boulder clay. The 'cave earth' layer was the second layer to contain artefacts, including lithic debitage and pottery. The flintwork was described as being either Late Mesolithic or Early Neolithic and includes a leaf-shaped arrowhead. The pottery is largely Peterborough Ware, although there are also some Beaker sherds (Gilks 1995, 4; Raistrick 1936, 193, 201).

Archive plans from the Pig Yard Club excavations show that the human bone from Sewell's Cave was found clustered together within a 1-metre area against the northern wall of the rock shelter (Leach 2006, 142). It was discovered at the intersection between the upper layer of clay with limestone blocks and the 'cave earth' deposit (Raistrick 1936, 193). The Sewell's Cave human bone was reassessed by Leach (2006, 2008) as part of the same study as the Thaw Head bone. She established (Leach 2006, 144) that there was no surviving post-cranial bone in the assemblage at all. There were a minimum of four individuals from the site: two adults, a child of about 3 years old and an infant of between 18 and 24 months. Individual 1 was a middle-aged man represented by fragments of the mandible, maxilla and cranial vault. A parietal bone from the cranium of this man was radiocarbon-dated

(OxA-13537: 5002 +/- 33 BP: see Appendix 1), giving a calibrated date range of between 3945 and 3700 BC at two standard deviations.

Leach's (2006, 144) study of the taphonomic evidence on the bone from Sewell's Cave showed that a few of the fragments had slight evidence for weathering, but there was no evidence at all of carnivore damage. The absence of any post-cranial bone led Leach (2008, 46) to suggest that the crania and mandibles had been brought to the site as the final stage of a secondary burial rite. If this were the case, then it would seem likely that during the earlier stages of the intermediary period, the corpses were buried, which would account for the absence of weathering and carnivore damage. In a similar manner to the Wataita example described in Chapter 3 (Kusimba et al. 2005, 247–250), the heads would then have been dug up and brought to Sewell's cave for their final burial. However, unlike in the Kenyan example, they were probably buried in a pit rather than being displayed in the rock-shelter entrance. This idea is supported by the evidence for the disturbance of the 'cave earth' layer at the point where the bones were found (Raistrick 1936, 193). The clustering of the bones within a 1-metre area and the relatively unweathered and ungnawed state of the bones (Leach 2006, 142, 144) also suggest that pit burial was the final act of the secondary rite in this case.

Early fourth-millennium BC burials in the Yorkshire Dales

The very similar early dates for the first burials at Kinsey Cave, Thaw Head Cave and Sewell's Cave might lead us to suppose that there was a regional cave burial rite early in the fourth millennium BC. Griffiths (2011, 1083–1084) has carried out a detailed review of the regional dating evidence and modelled the likely start of both the regional Neolithic and of cave burial in the region. Her conclusion is that it is *89.0% probable* that cave burial began before the Neolithic in North Yorkshire. As part of the same study, she also carried out new dietary stable isotope studies on the individuals dated by Leach (2006). All of the bodies from the Yorkshire Dales show an elevated level of nitrogen ($\delta^{15}N$ ‰) compared to animal bone from the same sites, suggesting a protein-rich diet (Griffiths 2011, 1080). This data includes both the very early burials discussed in this chapter but also the later 'fully Neolithic' individuals discussed in Chapter 6. As Table 5.2 shows, there was no significant chronological trend within this group of people. Hedges and colleagues (2008, 125–126) argue that a similar increase in $\delta^{15}N$ ‰ values in the assemblage from the

Table 5.2 Human $\delta^{15}N(‰)$ values for dated cave burials in North Yorkshire. Data from Griffiths (2011, 1101–1116).

Site	Lab Number	Date Range BC (2Σ)	$\delta^{15}N(‰)$
Kinsey Cave	OxA-15791	3970–3790	9.9
Thaw Head Cave	OxA-14264	3950–3715	9.8
Sewell's Cave	OxA-13537	3940–3700	10.0
Jubilee Cave	OxA-14262	3695–3530	10.3
Kinsey Cave	SUERC-10518	3660–3520	8.7
Cave Ha 3	OxA-13539	3655–3520	13.7
Lesser Kelcoe Cave	OxA-13538	3650–3520	10.9
Cave Ha 3	OxA-14266	3515–3110	10.5

Hazleton North chambered tomb is part of a pattern of generally elevated nitrogen values for all Neolithic sites, indicating a diet rich in animal protein in the period. Griffiths' (2011, 1080) modelled estimate shows that, for the people buried in North Yorkshire caves, this highly carnivorous diet was well established very early in the fourth millennium BC.

All three sites seem to have been used at the same time, and the people buried within them had very similar diets. Long time-scales are necessary to develop the changes in bone chemistry being measured by stable isotope analysis; typically, they average diet over a period of years (Hedges et al. 2008, 116). Therefore, it is probable that the people buried in these three caves shared a similar set of relations with their environment. They may have been hunter-gatherers with a way of life based on large mammal hunting or highly carnivorous pastoralists, of the kind Mlekuž (2005, 29–34) has suggested inhabited the Eastern Adriatic region before 4800 BC. However, despite these similarities, when these people began to use caves for extended burial rites, they did so in different ways. In both of the North Yorkshire examples, where there is well-dated evidence of early burial, the idea of an intermediary period seems to have been important. At Sewell's Cave, I think that we can see clear evidence of a secondary burial rite focussed on the head. By contrast, the evidence from Thaw Head Cave shows an extended and complex intermediary period for bodies which had been successively inhumed within the cave. In this case, the bodies did not move locations during the intermediary period, but the agency of both cave processes and animals played a large part in the transformations necessary to mark the stages of the intermediary period.

Burials and society in transition

It was suggested by Hellewell and Milner (2011, 62) that burial which involved an intermediary period and the fragmentation of the body began in the Late Mesolithic. As I have shown in this chapter, evidence for this is extremely partial and only preserved in a few places. The Oronsay shell middens provide the best evidence for a multi-stage burial rite which was definitely being practiced by hunting and fishing communities (Meiklejohn et al. 2005, 89–96). However, it must be noted that most of the documented midden burials in Scotland are of Neolithic or later date (Armit and Finlayson 1992, 669). Meiklejohn and colleagues (2005, 100–101) also pointed out that the Oronsay shell-midden burials were not obviously similar to Mesolithic burials in either southern Scandinavia or Brittany.

As shown by the examples in this chapter, it is highly probable that some multi-stage cave burials took place in what is currently considered to be 'chronologically' the Late Mesolithic. At George Rock Shelter and in the Yorkshire Dales, we have sites which were used so early that they are very likely to have pre-dated the local modelled estimates for the start of the Neolithic. There are two possible explanations for this phenomenon. Either cave burial was first practiced by hunter-gatherers, as Hellewell and Milner (2011, 62) suggest, and there was continuity of practice into the Neolithic. This would imply a piecemeal adoption of some elements of Neolithic practice and at least some continuity of population. The best support for this comes from the shell-midden sites of western Scotland. Alternatively, cave burial may have been a solely Neolithic practice, as Schulting and colleagues (2013, 22) suggest. However, because it was not usually associated with diagnostic material culture, these sites have not been used in the regional models for the start of the period. This may have resulted in a slight mis-dating of the start of the Neolithic, giving the impression that the cave burials are too early to belong in the period. This explanation would fit better with currently developing models which stress a decisive break between the Mesolithic and Neolithic and a substantial population change at this time (Brace et al., in preparation). It should be noted that neither explanation covers all the evidence presented in this chapter, and it is likely that elements of both apply to different regions to different degrees. However, as I will explore in more detail in Chapter 6, there is much more cave-burial evidence from the Early Neolithic. Early fourth millennium BC cave burial was clearly part of a developing set of ideas about death, time and transformation which went on to be extremely influential in the Early Neolithic.

As described in Chapter 2, early fourth-millennium cave burials in continental Europe were all being carried out by people living a Neolithic way of life. Some similar burial practices were in use at these caves as in the early British sites. For example, Abri des Autours, Namur, appears to have been used for secondary burial between 4320 and 3980 BC. This site also had evidence of the use of drystone walling to enclose burials (Polet and Cauwe 2007, 74–84). Secondary burial was also taking place at Höhlenstein-Stadel, Baden-Württemberg, between 4470 and 4040 BC (Orschiedt 2012, 218). Successive inhumation burials took place at Les Grottes des Barbilloux, Lot-et-Garonne and L'Abri du Pas-Estret, Dordogne, between approximately 4500 and 3700 BC (Beyneix 2012, 225–226).

It may be that the adoption of multi-stage burial rites at natural locations, and particularly at caves, was something that Late Mesolithic people did. However, if they had learnt these practices from contact with farming groups in Europe, it could be argued that multi-stage cave burial was actually the earliest element of a Neolithic way of life to be adopted. This may have occurred, at least in North Yorkshire, as much as a century (Griffiths 2011, Figure A1.13 and see Figure 5.6) before the introduction of other, more obviously Neolithic, things such as monuments, pottery or domesticated plants and animals. What is clear is that whether multi-stage cave burial is regarded as culturally 'Mesolithic' or culturally 'Neolithic', then common ideas about death, physical and social transformations and human remains were shared between Britain and the continent from the beginning of the fourth millennium BC.

6

Written on the body

Introduction

In the previous chapter, I established that multi-stage cave burial rites in Britain had their origins at the beginning of the fourth millennium BC. The review of this evidence also shows that even within these few early cave burials, there were different rites and practices. In this chapter, I will discuss the greater range of evidence we have for cave burial from after around 3800 BC. By this date, in all of the regions of Britain where there are cave burials, a Neolithic way of life was at least a possibility. This is not to say that everyone who was buried in a cave after 3800 BC should automatically be assumed to have been a farmer or a pastoralist, but, by this date, it is reasonable to assume that cave burial would have been taking place within a broadly 'Neolithic' culture. The diversity of cave burial rites which existed early in the fourth millennium seems to have continued into the Neolithic period. These different rites included various kinds of multi-stage funeral, all of which presumably incorporated an intermediary period. To return to the terminology established in Chapter 3, we can see evidence for both secondary burial and successive inhumation. Within the broader category of secondary burial, there were a range of different possibilities, depending on where the intermediary period took place and which people, animals and natural phenomena acted on the body during this time. A few caves also seem to have been used for primary burial without any intermediary period. Figure 6.1 shows the locations of all the cave sites discussed in this chapter.

A cult of the head

The early fourth millennium BC burials from Sewell's Cave, North Yorkshire, discussed in the last chapter appeared to represent a highly distinctive form of funeral in which only the head was chosen for

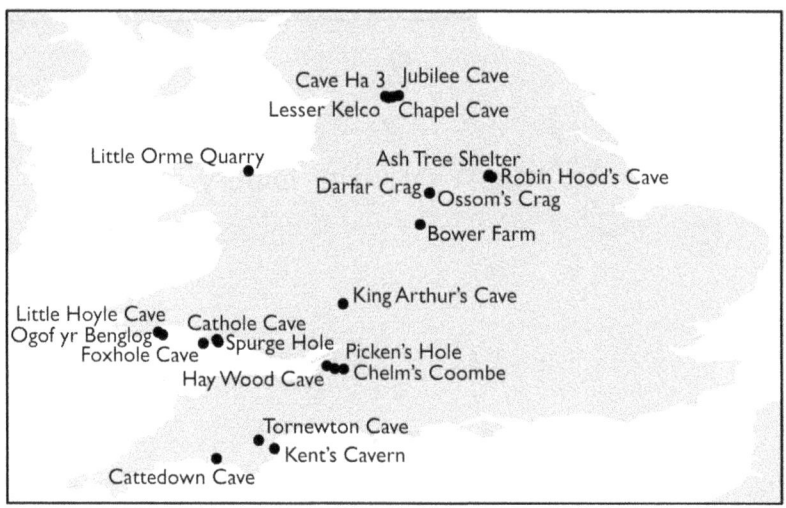

6.1 Location map for the sites discussed in Chapter 6. The base mapping includes data licenced from © EuroGeographics.

the final secondary burial. This rite can be identified at two other, slightly later, sites in the Pennines: Robin Hood's Cave, Derbyshire; and Lesser Kelco Cave, North Yorkshire. A similar rite focussed on the collection or curation of heads has sometimes been suggested for other cave sites, for example the rock shelter at Bower Farm, Staffordshire (Cane and Cane 1986, 3), where the only human bone identified by the excavators were parts of two adult female crania. However, further study of this assemblage identified post-cranial bone which has been radiocarbon-dated to the Early Neolithic (Meiklejohn et al. 2011, 34). This indicates a wider problem. Crania in particular are both highly visible and easily recognisable by non-specialists as human bone. Therefore, they tend to be over-represented in earlier excavation accounts. There may be more Neolithic examples of secondary burial of the head than I have identified here, but they do not have such unambiguous taphonomic and skeletal evidence for the rite.

Lesser Kelco Cave, North Yorkshire (Appendix 1, number 27: NGR SD 8098 6467), is another site on Giggleswick Scar, south-east of Kinsey Cave. It was excavated between 1928 and 1932 (Simpson 1950, 260–261). The deposits in the cave were approximately 3.3 metres deep at the entrance (see Figure 6.2), and Simpson (1950, 260–261) identified four main stratigraphic events in the fill. The uppermost of these was a layer of fallen stalagmite and breccia mixed with limestone fragments, which was around 0.3 metres thick. This sealed a cave earth layer which Simson's section drawing suggests was around 1.8 metres deep. All of the archaeological finds came from these two uppermost layers. Apart from the human bone, there were faunal remains from both wild and domesticated species, charcoal spreads showing the former position of hearths and Romano-British and Middle Neolithic pottery sherds. The positions of the finds were documented by Simpson (see Figure 6.2). It is likely that the crania were originally buried into the Neolithic surface of the cave floor.

Four fragmentary crania were recovered by Simpson's team, and a fifth was apparently removed by schoolboys after the end of the 1930 season. The four crania were studied in 1933 by Dr Cameron of the Royal College of Surgeons, who identified two adult women, an adult man and an adolescent. The human bone from the site is now in the Lord collection at Lower Winskill, where it has recently been re-assessed by Leach (2006, 185–187). She identified a minimum of three individuals from the site, based on the surviving cranial fragments: one adult female, equivalent to 'skull 1' in Simpson's (1950, 262) report; and two adult males. She also identified two post-cranial bones from the site: a left ulna and a right humerus (Leach 2006, 187). Despite the presence of these two post-cranial bones, Leach considered that Lesser Kelco Cave was another site where the secondary burial of heads had taken place (Leach 2008, 51). The two arm bones were found at a

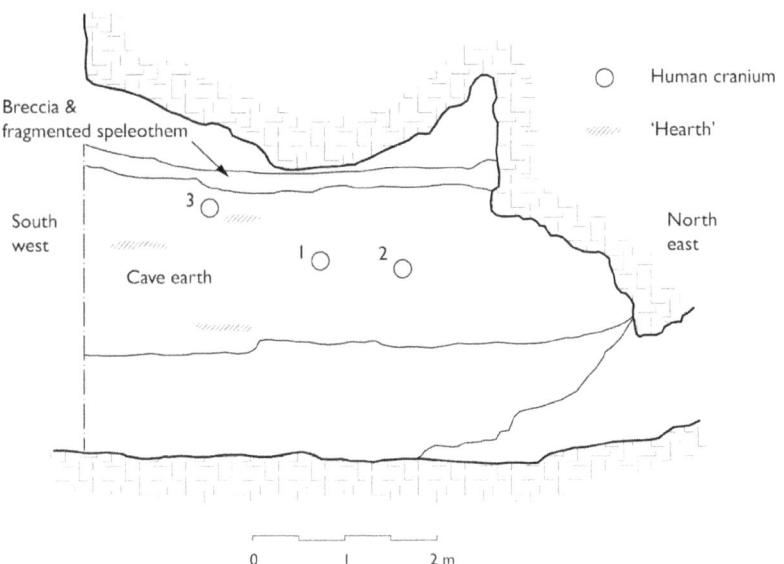

6.2 Section through the excavated deposits at Lesser Kelco Cave based on Simpson (1950, 261) and with additional information from Leach (2006, 187).

higher level of the cave deposits than the crania and nearer to the entrance. They also had taphonomic evidence of extensive carnivore attrition (Leach 2006, 187). She considered that these two elements were probably introduced into the upper levels of the cave at a later date by scavenging animals (Leach 2006, 189).

The surviving cranial bones had a completely contrasting set of taphonomic signatures. Like the Sewell's Cave bones, they had only slight evidence of weathering and no sign of carnivore scavenging (Leach 2006, 190). The single radiocarbon date from the site (Appendix 1) comes from a bone in skull 1. This date (OxA-13538: 4801 +/- 31 BP) would give a calibrated range of 3650–3520 BC at two standard deviations. Griffiths' (2011, 1080 and see Table 5.2) study of the stable isotope values from North Yorkshire cave sites shows that this woman would have shared the meat-dominated diet of the other burials in the region. The three identifiable individuals in Lesser Kelco Cave seem to have had a very similar funeral to the people buried in Sewell's Cave. After death, their bodies were buried for possibly as long as a few years. The burial sites would have been known or marked so that, once the intermediary period was over, the crania could be removed and brought for final burial in the cave. The recorded locations of the excavated skulls might show that, unlike the Sewell's Cave examples, each head at Lesser Kelco was buried in a separate pit.

Robin Hood's Cave, Creswell Crags, Derbyshire (appendix 1, number 42: NGR SK 5341 7419), is a cave with a long history of excavation (Campbell 1977, 64–65). The Neolithic radiocarbon dates came from the fragmented remains of what was probably a single individual discovered during Campbell's excavations at the site in 1969. Campbell recovered eight bone fragments, but only one of these, part of a frontal bone, came from an undisturbed layer. This find came from a scree deposit outside the cave entrance and it gives some indication where the bone was originally placed. The remaining fragments were all recovered in the spoil from nineteenth-century excavations in the same area (Campbell 1977, 69 and see Figure 6.3).

The interpretation of the Robin Hood's Cave burial as a secondary head burial depends primarily on osteological analysis undertaken by Powers and Campbell and reported as Appendix 6 of Campbell (1977, 218–220). At the time this report was written, the bones were assumed to date to the Late Upper Palaeolithic. The eight bone fragments they identified included five cranial elements, two fragments of mandible and a single third cervical vertebra (see Figure 6.4). They argued that this assemblage was the remains of a head which had been severed below the third cervical vertebra and brought to Robin Hood's Cave as a trophy (Campbell 1977, 219). While the

6.3 The excavated area at Robin Hood's Cave and the location of the human frontal bone fragment 465 and the other cranial and vertebra fragments (after Campbell 1977, Figures 27 and 35 and Appendix 6).

head-hunting hypothesis may be the correct explanation, a secondary head burial with a short enough intermediary period that the cervical vertebra and mandible retained some connecting tissues could also have created this assemblage of bones. The diagnostic factor here is likely to be the cervical vertebra, as these tend to disarticulate early in a decomposition sequence, whereas the joint between the cranium and mandible disarticulates later (Knüsel 2014, 32–35). It is notable that the Sewell's Cave secondary head burials included mandibles but not cervical vertebrae (Leach 2006, 144), which might tend to suggest a slightly longer intermediary period at that site.

The fourth-millennium BC dates from Robin Hood's Cave (see Appendix 1) are from the frontal bone fragment (number 465: OxA-7386: 5000 +/- 40 BP) and the vertebra (number 132: OxA-1807: 4870

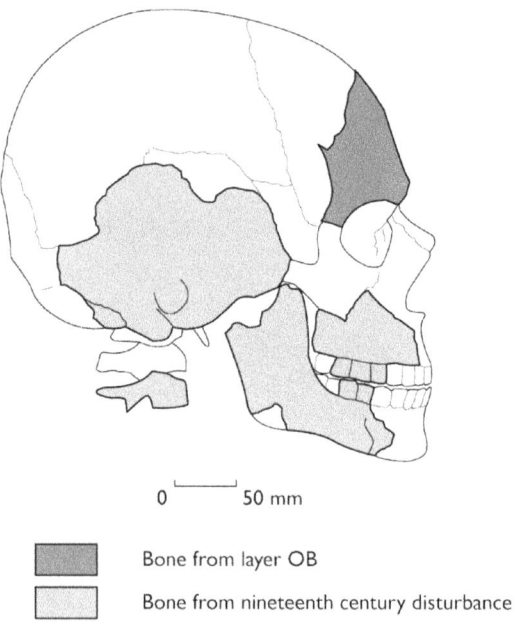

6.4 The surviving elements of the Robin Hood's Cave head as reconstructed by Powers and Campbell (after Campbell 1977, 325).

+/- 120 BP). These results are statistically consistent and potentially could date a single individual (Griffiths 2011, 862). However, when Hedges and colleagues (1991, 290–291) reported the result on the vertebra, it was described as a 'lumbar vertebra'. This identification has since persisted in the literature, for example, Charles and Jacobi (1994, 17) and Griffiths (2011, 894). If this more recent identification of the vertebra is correct, then this significantly decreases the likelihood that the Robin Hood's Cave bones are the remains of either a trophy head or a secondary head burial. I have not been able to trace any published discussion of the re-description of this element and have provisionally accepted the original identification made by Powers and Campbell (Campbell 1977, 219).

If Powers and Campbell (Campbell 1977, 219–220) were right in their original interpretation of the Robin Hood's Cave material as the remains of a trophy head, then this may be relevant to secondary burial in a slightly different way. Ethnographic accounts of secondary burial in Indonesia collated by Hertz (1960, 201) provide several instances where one of the conditions which had to be met to mark the end of the intermediary period in a secondary burial was the successful taking of a head by a relative of the deceased. Without postulating a direct analogy, it is possible that successful raiding or inter-group violence was a way of marking the end of the intermediary period. Schulting and Wysocki (2005, 128–129), in their broader review of the evidence for inter-personal violence in the period, provide a list of crania with associated cervical vertebra from other kinds of Neolithic site. They list six possible examples and make the excellent point that a group which afforded secondary burial to the head, and therefore probably regarded the head as powerful, would be likely to believe that taking the head of an enemy would disrupt and appropriate their power. However, Schulting and Wysocki (2005, 129) were unable to locate any evidence for cut marks on cervical vertebrae or mandibles to definitively identify deliberate removal of any of the heads they listed, and Leach (2006, 149, 190) notes a similar absence of cut marks on the Sewell's Cave and Lesser Kelco Cave examples.

Evidence for similar rites around either the collection or secondary burial of heads is comparatively rare from sites in continental Europe and much earlier than the British evidence. Robb (2007, 58–60) cites examples from the Italian Neolithic, a particularly well-documented example coming from Grotta Scaloria, Puglia, in the second half of the sixth millennium BC (Robb et al. 2015, 42 and see Chapter 2). During the Mesolithic period in Alsace and southern Germany, a group of seventh and early sixth-millennium BC caves have deposits of crania with articulated mandibles and cervical vertebra. Remains at these sites do have cut marks to the vertebrae and also often show signs of blunt-force trauma, strengthening the argument that they were trophy heads collected in conflicts (Orschiedt 2012, 215).

Sewell's Cave, Lesser Kelco Cave and Robin Hood's Cave all provide good evidence for a keen Late Mesolithic and Early Neolithic interest in the human head. This took the form of a specific secondary head burial rite and may also – suggested more tentatively – have involved the collection of trophy heads during raids. There is also possible evidence of this head cult at long barrows such as Bole's Barrow and causewayed enclosures such as Staines (Schulting and Wysocki 2005, 128). In this rite, living people and the decomposing corpse acted together to create the temporality of the practice. The timings of the human interventions in the decomposition processes would have had to have been carefully judged to both create and curate the separated cranium in an appropriate way. The directionality of the decomposition process and the way it would have indexed the passage of time might be thought of as the kind of entangled relationship between people and things described by Hodder (2012). Caves, in contrast to the sites of some of the other rites considered in this book, would have provided a relatively passive container for the final secondary burial. Reilly (2003) makes a somewhat similar argument for the Orcadian Earlier Neolithic. He argues that dead bodies in chambered tombs on Rousay were initially placed in the chambers of tombs on the lowest terrace of the island. During the intermediary period, as the body reached a certain stage of decomposition, the major long bones and crania were moved into monuments on the upper terraces. This process ended with the collection and arrangement of disarticulated crania in the least accessible chambers of these monuments. Reilly (2003, 140–143) argues that the whole island functioned as a locale for a secondary burial rite which manipulated the natural processes of bodily decomposition to 'distil' the essence of the dead body into a cranium.

Mummification and curation

The interaction between living people and the decomposition of a corpse was one of the ways that Hertz (1960, 201) expanded his definition of the intermediary period to include practices such as mummification. The decomposing body may be one kind of participant, defining and creating a particular tempo of change, but the mourners and embalmers create through their actions a different temporality, usually a much longer-term one. The preserved body may become part of a delayed primary burial rite, as with the Philippine examples considered in Chapter 3, but it may also remain for much longer within the daily experience of the living. To maintain its mummified state, the corpse may need to be looked after or even repaired. As long as the body is being curated and tended, it will have its own embodied narrative indexed by the physical traces left by the mummification and repair. The evidence for mummification in prehistoric Britain has been recently reviewed by Booth and colleagues (2015). In this paper, they argue that the consistent effect of all types of mummification is to prevent or greatly reduce the bacterial bioerosion of bone. In cases of burial without mummification, putrefying bacteria originating in the gut flora attack both soft tissues and the internal structure of the bone. This damage is detectable in microscopic cross-section. Booth and colleagues (2015, 1161–1163) were able to demonstrate, from a sample of 301 prehistoric individuals of all periods, that it was highly probable that deliberate mummification took place for 16 of the 34 bodies from Bronze Age sites in Britain. No other prehistoric period had this evidence, and they therefore argued that mummification was a particularly Bronze Age practice (Booth et al. 2015, 1163). Unfortunately, as the putrefying bacteria originate in the gut flora, then bodies which have not been buried may have low levels of bacterial bioerosion even if they have not been mummified. This is because other processes, such as scavenging or invertebrate action, may consume the soft tissue before bacterial bioerosion of the bone is far advanced (Booth et al. 2015, 1161). The Neolithic individuals in this study predominately came from disarticulated skeletons, and therefore Booth and colleagues (2015, 1163) were not able to rule out the possibility of some Neolithic mummification having taken place.

Detecting the presence of mummification from Neolithic caves is therefore likely to rely on other types of evidence. It is necessary to look at the embodied narrative of change involved in mummification, what Lemonnier (2012, 16–17) would describe as the 'cultural technology' of mummification. The two most likely methods for mummifying the dead in British prehistory are either through temporary immersion in a peat bog (suggested by Parker Pearson and colleagues [2005, 542] for the mummies from Cladh Hallan, South Uist) or by smoking the corpse (suggested by Booth and colleagues [2015, 1169] for the individual from Neate's Court, Kent). In both cases, the body would have to have been eviscerated beforehand to remove the major source of putrefying bacteria. Drawing on the example of the Later Bronze Age mummies at Cladh Hallan, then it is clear that two other stages of the mummification process may also leave traces which can be detected archaeologically. The mummies discovered beneath the north house at Cladh Hallan were in extremely tightly flexed postures, and it was suggested that this was evidence that they were wrapped to create a 'mummy bundle' of the kind from the central Philippines discussed in Chapter 3. Both bodies had also been modified or curated after mummification. One burial was a composite made up of the cranium and cervical vertebrae from one individual, the mandible of another and the post-cranial skeleton of a third (Parker Pearson et al. 2005, 534–535).

Human remains from Spurge Hole Cave, Gower (Appendix 1, number 45: NGR SS 5468 8730), may provide evidence for mummification in a Neolithic cave burial. The cave entrance is a small arch 1.2 metres wide, situated part-way down a sea cliff on the south coast of the Gower peninsula in south-west Wales. It was discovered and excavated by Mel Davies in March 1985. He recorded an extended adult inhumation beneath a gravelly deposit in the cave entrance (Davies 1989a, 88). There was a reinvestigation of the site in 1991 by a team from the National Museum of Wales. They recovered the human remains from around 0.25 metres down in the entrance deposits. They were able to confirm that the burial was extended in an east-west orientation with the head to the west across the cave entrance. Osteological analysis of the bones showed that the apparent individual burial was a composite made up of at least two individuals. The right pelvis and some cranial fragments of the surviving bone can be attributed to a male adult, while the left pelvis and left and right femurs belonged to an adult female. There was at least one further individual in the cave represented by a juvenile tooth (National Trust HBSMR 2003). The radiocarbon date (OxA-3815: 4830+/-100 BP: Appendix 1) comes from the left femur and therefore dates the female part of the possible mummy. The date for the Spurge Hole burial is unsatisfactory from a purely chronological point of view for a number of reasons. It is now some time since the measurement was carried out, and the large standard deviation makes the calibrated date intrinsically imprecise when compared to more recent radiocarbon results. However, there is also the problem that we do not know what the chronological relationship was between the parts of the composite burial and, if the bodies were mummified or otherwise curated, how long after the death of this woman the composite burial took place.

The suggestion that the Spurge Hole burial is the remains of a Neolithic mummy must remain very tentative. It is solely based on the facts that the excavation report described the burial as an extended single individual and that the osteological data indicated that this individual was made up of more than one person. Other explanations are entirely possible. The Spurge Hole composite burial may have been created not by the creation and repair of a mummified body, but by the deliberate arrangement of skeletal elements as part of a secondary burial. Similar attempts to create 'individuals' from skeletonised fragments in Early Neolithic collective deposits were noted by Wysocki (Wysocki and Whittle 2000, 598) in his analysis of the remains from the Penywyrlod, Pipton and Ty Isaf chambered tombs, although, in these cases, the process seems to have been less complete than at Spurge Hole.

Mummification and body curation is known from some European caves. Antiquarian excavations recovered mummified bodies accompanied by preserved organic clothing and shoes, which have subsequently been dated to the early fifth millennium BC, from Cueva de los Murciélagos, Granada (Weiss-Krejci 2012, 127). There is also evidence for human bone having been curated for several centuries from four Chalcolithic cave sites in the Iberian Peninsula, which may indicate that mummification was a longer-lasting practice there. Casa da Moura, Gruta do Cadaval, Gruta dos Ossos and Covão d'Almeida all have fifth and fourth-millennium BC dates for human bone from deposits which are otherwise securely dated to the mid to late third millennium BC (Weiss-Krejci 2012, 130).

Secondary burial

All of the funerary rites considered so far in this chapter are likely to have involved an extended intermediary period and therefore should be considered as special cases of secondary burial. The manipulation of hands and feet identified by Meiklejohn and colleagues (2005, 102–103) at Cnoc Coig and Carding Mill Bay 1 was discussed in Chapter 5. This would also have required mourners to monitor the changing condition of the decaying bodies closely and to move these body parts at least from one part of the midden to another and possibly further. These rites would certainly have incorporated an intermediary period, but it is not clear to what extent the skeletal remains were being moved from one location to another and therefore whether we should properly refer to them as secondary burials. There are other cave sites where the evidence suggests that a secondary burial rite was in use. As discussed in the last chapter, the bone assemblage at George Rock Shelter suggests that site was a place where intermediary-period burial happened. The tooth-dominated assemblage at Broken Cavern might indicate that this was another cave used for intermediary-period burial.

Chelm's Combe, Somerset (Appendix 1, number 17: NGR ST 4634 5447), was a medium-sized rock shelter, 9 metres long and up to 6 metres deep, in the south end of Old Chelmscombe Quarry, Cheddar Gorge, which has now been quarried away (Figure 6.5). The site was excavated by the Somerset Archaeological and Natural History Society in 1925 under the on-site supervision of the highly skilled professional archaeologist W.E.V. Young (Balch and Palmer 1926, 97–100). The site is unusual: in the cliff face below the main rock shelter, covered by limestone scree, the excavation team discovered a small rock-cut chamber around 0.9 m wide and similarly deep, which had been used for burial. There were also large quantities of human remains discovered in the upper fills of the main rock shelter. The rock-cut chamber is the only example I am aware of from a British site where an artificial cave, albeit a very small one, has been created in limestone. As discussed in Chapter 2, rock-cut tombs were a major class of collective burial site in Mediterranean limestone regions, particularly in the later fourth and early third millennia BC. In these regions, they were closely associated with natural burial caves.

Young appears to have dug the site using a system of measured 1-foot-deep spits. He probably adopted this approach from the contemporary excavations at Windmill Hill, Wiltshire, where he was also being employed in the mid-1920s. Finds from the site are therefore reported by their depth from the surface. A more general account of

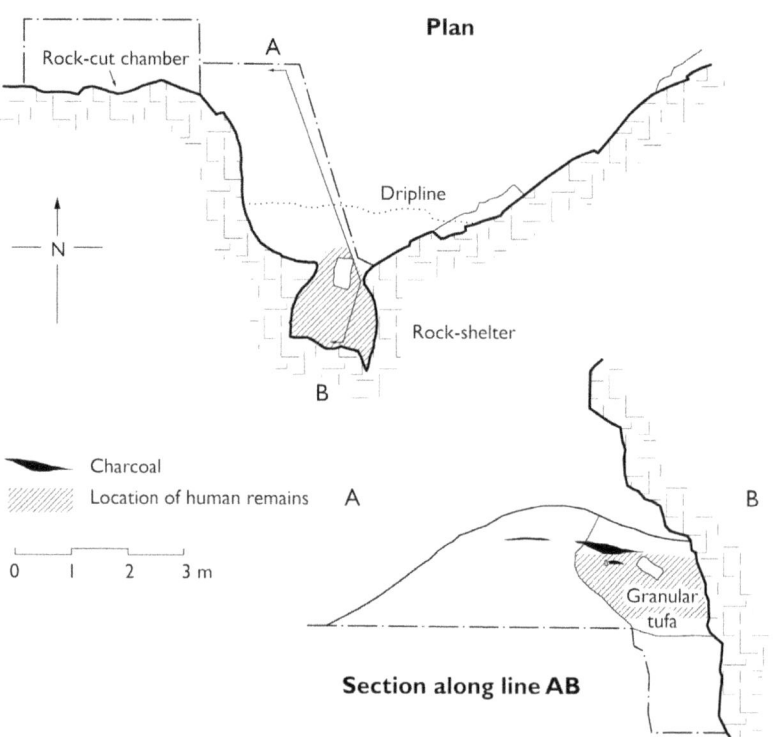

6.5 Plan and section of the excavated deposits at Chelm's Combe (after Balch and Palmer 1926, Figure 3).

the stratigraphy of the cave is also given in the report. A modern soil up to 1 metre in depth covered a granular tufaceous deposit inside the shelter. This deposit was around 3 metres deep. There were faunal remains of glacial species such as reindeer in the lower parts of this deposit, and therefore it presumably formed gradually from the beginning of the Holocene (Balch and Palmer 1926, 98). The excavation continued for another 4.8 metres, all the finds from these lower layers being Pleistocene faunal remains. The skeletal material from the site was catalogued by Cooper (Balch and Palmer 1926, 101–106). It is presented in the report as a full list of individual elements. From this catalogue, it is possible to identify the number and type of bones surviving from each layer of the rock shelter and from the rock-cut chamber. The deepest surviving human bone from the rock shelter came from around 2.7 metres into the granular tufa deposit, and the bulk of it was discovered between 1 and 1.5 metres into this layer. There were also sherds of Neolithic pottery within this part of the granular tufa deposit. At around 1.2 metres down were two relatively complete Early Neolithic bowls, together with fragments of at least four other vessels of similar date; one was a lugged Hembury-style bowl and the other a decorated Windmill Hill–style vessel (Balch and Palmer 1926, 108–110). There were also eight flint scrapers from this level. However, the presence of Beaker and Peterborough Ware sherds in the assemblage indicates that this was not a closed context of a single date but an open deposit which developed gradually. The single radiocarbon date from Chelm's Combe was obtained on a long bone from either the rock-cut chamber or the rock-shelter levels (BM-2974: 4680 +/- 45 BP: Appendix 1) and would calibrate to between 3630 and 3365 BC at two standard deviations.

The first point of interest in attempting to understand the funerary rite at Chelm's Combe is that the bone from the rock-cut chamber was much better preserved than that from the rock shelter. The only intact crania came from the chamber, and Cooper noted the generally much more complete preservation of individual elements from this part of the site (Balch and Palmer 1926, 102–104). Cooper calculated that there was a minimum of five individuals buried in the chamber, and it is noticeable (see Figure 6.6) that they were overwhelmingly represented by the bones of the head, trunk and major limbs. By

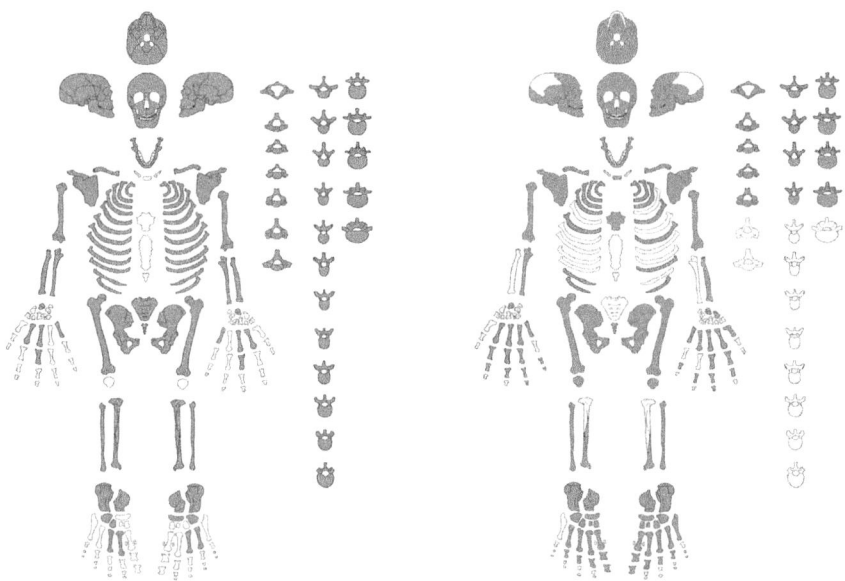

6.6 Surviving elements in the Chelm's Combe skeletal assemblages from (left) the rock-cut chamber and (right) the granular tufa deposit in the rock shelter showing the different taphonomic signatures for each area. Based on data in Balch and Palmer (1926, 101–106).

contrast, the fragmentary remains from the rock shelter included large numbers of disarticulated teeth, hand and foot bones and elements such as patellae and the hyoid. Based on these observations, I would suggest that there were two different stages of a secondary burial rite at Chelm's Combe. During the intermediary period, the bodies would have been placed in the rock shelter, among the accumulating granular tufa and scree, in a somewhat similar way to the example from George Rock Shelter discussed in Chapter 5. These bodies may have been accompanied by pottery or, perhaps more plausibly, by food contained in pottery. After a relatively protracted intermediary period, long enough for some of the crania to fragment into the separate bones of the skull, the surviving long bones and crania were moved to a different location for secondary burial. It is extremely tempting to think that, in this case, the location for the secondary burial was the nearby rock-cut chamber. The bone assemblage from that part of the site seems to represent a secondary burial assemblage, dominated by disarticulated crania, long bones and axial elements (see Figure 6.6). Cooper also noted evidence of carnivore damage on two of the femurs (Balch and Palmer 1926, 104), which is further evidence that they were exposed prior to their final burial.

Secondary burials of the kind discussed in this section are known from a wide range of cave sites outside of Britain. In Ireland, Dowd (2015, 9–100) notes evidence of the use of Annagh Cave, Limerick, for both the intermediary period exposure and final secondary burial of bodies. In the Belgian Middle Neolithic, the most common cave burial rite appears to be successive inhumation, but there is good evidence for a secondary burial rite at Abri des Autours, Namur (Polet and Cauwe 2007, 74–84). Both Jungfernhöhle, Bavaria; and Höhlenstein-Stadel, Baden-Württemberg, were being used for secondary burial in the late fifth and early fourth millennia BC (Orschiedt 2012, 217–218). Evidence for secondary burial in France, at Can-Pey, Pyrénées-Orientales (Baills and Chaddaoui 1996, 367), is not very precisely dated but is later, in the late fourth or early third millennia BC.

Secondary burial rites did take place in British caves. However, this review has highlighted that there was not a single secondary burial rite for caves. The physical and social changes which created the specific temporality of the intermediary period, the need to deal with incomplete exchanges, grief, unpaid obligations and bodily decomposition, were responded to in different ways. In some places, as for example with the midden burials, then it seems that the important contribution of the cave environment was to provide a space which physically indexed the long-term passage of time. These sites, with their established shell middens and accumulations of artefacts, would have provided a *circulating reference*, in Latour's (1999, 69–79) terms. This would have linked the ongoing temporal processes around burial and decomposition with indices of much older changes within the cave. In others, such as the secondary head burials, the temporality of the intermediary period was experienced away from caves. Here, the intermediary period could be seen as primarily driven by human interventions in the processes of bodily decomposition. This rite would have ultimately created an artefact, the separated head, which indexed the whole complex of beliefs and practices around death. Caves were then chosen as the appropriate place to bury this extremely powerful object. There was also the possibility, as with the Chelm's Combe example discussed here, that all stages of the secondary burial process took place within a single cave or complex of caves. In these cases, references and indices may have been distributed over the nearby landscape so that the burial process drew upon and was constituted though changes to the whole environment rather than to specific caves. I will return to this possibility in more detail in Chapter 8.

Primary burials

Individual burials do exist from the British Early and Middle Neolithic, despite the emphasis in the published literature on collective deposits. Schulting (2007, 583–584) has reviewed the evidence for primary burials from otherwise unmarked flat graves in the Neolithic. Early to Middle Neolithic examples include three at Barrow Hills, Oxfordshire; and two at the Eton Rowing Course site, Buckinghamshire. There are two examples of Earlier Neolithic cave sites with what may be primary burials, in the sense the term is used in Chapter 3 and by Knüsel (2014, 46).

Jubilee Cave, North Yorkshire (Appendix 1, number 23: NGR SD 8376 6551) is a small passage cave at the north end of Attermire Scar. The site has a complicated excavation history, but the most recent excavations were carried out by Tot Lord and Arthur Raistrick between 1935 and 1940. The records and finds from this work are now curated in the Lord archive at Lower Winskill (Leach 2008, 41). There are two parallel phreatic passages around 8 metres long which terminate in a small chamber. The majority of the human bone from the site was discovered, apparently articulated, beneath a rock ledge at the back of a side fissure (see Figure 6.7), with a few fragmentary pieces also discovered closer to the main passage. There are a range of finds of different dates from the cave, including Mesolithic flintwork, Peterborough Ware and Romano-British pottery (Leach 2006, 193–194). There are a minimum of five individuals represented in the

6.7 View along the east fissure of Jubilee Cave towards the rock ledge where the burial was discovered.

skeletal assemblage. However, the vast majority of the bone comes from a single man, individual 1. The actual figure recorded by Leach (2006, 195) is that 74% of the assemblage is identifiable as being part of this body. However, this does not include the cranium and mandible, which were recovered during the excavations but subsequently lost. This body seems to have been a primary burial beneath the rock ledge, although the extremely fragmentary remains of the other four individuals obviously show that there must have been other funerary rites taking place in the cave at some date (Leach 2006, 200–201).

Leach's (2006, 195) assessment of the human remains from individual 1 shows that, in contrast to the other fragments, it was not weathered and did not show any evidence of vertebrate scavenging. The only date from the site comes from the tibia of this man (Appendix 1: OxA-14262: 4836 +/- 31 BP). This would calibrate at two standard deviations to between 3695 and 3530 BC. It is clearly possible that the fragmentary remains at Jubilee Cave represent earlier burials in a phase of successive inhumations, although, as they do not have radiocarbon dates, their precise relationship to individual 1 is unclear. However, the distinctively different level of preservation on the adult male inhumation does suggest that this burial was an example of a primary burial. The taphonomic signature at Jubilee Cave can be contrasted with the much more fragmented individuals from nearby Thaw Head cave (see Chapter 5). Both sites have a similar number of individuals, but the Thaw Head assemblage appears much more consistent and suggests that the same successive inhumation rite was used for each burial.

The human remains from Little Orme Quarry, North Wales (Appendix 1, number 29: NGR SH 8176 8248), were found during nineteenth-century quarrying within the fill of a widened fissure in the limestone. The skeletal material was around 15 metres deep in the fissure, which was exposed in section by the quarry (Gregory et al. 2000, 3–4 and see Figure 6.8). The human remains were fully described by Gregory and colleagues (2000, 5–6), and it is clear that they represent the reasonably well-preserved remains of a single individual. Parts of all the major elements of the skeleton were present, included extremities such as hand and foot bones. The skeleton was of a woman who was exceptionally old, being somewhere between 54 and 63 years old at the time of her death.

The radiocarbon date for this woman comes from a portion of femur (Appendix 1: Beta-87306: 4720 +/- 50 BP). At two standard deviations, this date would calibrate to between 3640 and 3360 BC. This burial is likely to have been a primary burial placed into the partially filled fissure within the limestone. There was a Late Bronze

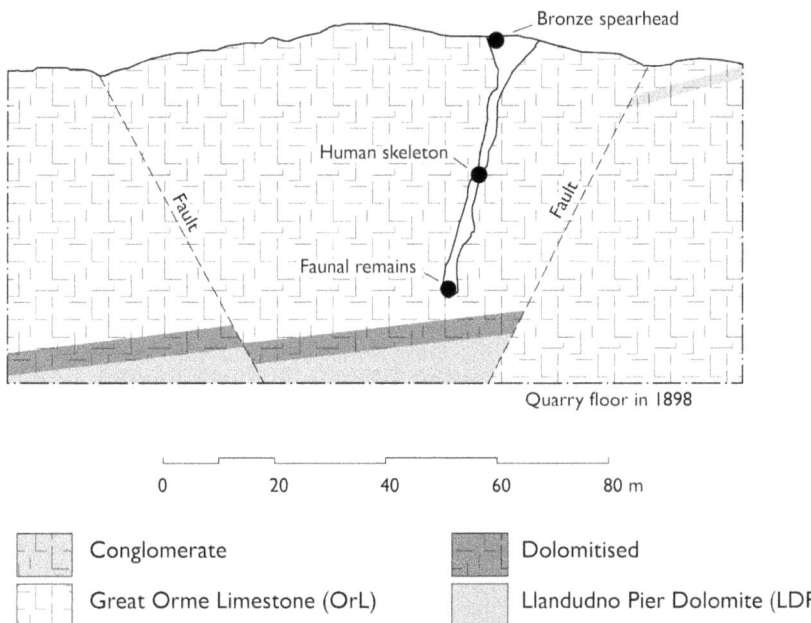

6.8 The reconstructed location of the Little Orme Quarry human remains (after Gregory et al. 2000, Figure 2).

Age socketed spearhead discovered in the upper fissure fill, just below the nineteenth-century ground surface (Gregory et al. 2000, 6–7). This implies that the fissure continued to fill naturally over this period.

Distinguishing between primary burials and successive inhumation sites with low overall numbers of burials has been problematic. The rites ought to be distinct, as successive inhumation implies an intermediary period during which the mourners, the decomposition processes and the environment can act. In practice, the active nature of cave deposits means that even a burial which was intended to be a primary burial may have some of the traits of a successive inhumation. In areas where primary burials are regarded as the norm in caves, such as the Early Neolithic examples from Italy, Southern France and Western Spain discussed in Chapter 2, then researchers have noted that the apparently standard primary burials conceal a greater diversity of practice (Robb 2007, 57–60; Zemour 2011, 261). In a similar way, the apparent preponderance of successive inhumation sites in British Early Neolithic caves probably includes some further unrecognised examples of primary burial. Primary burial is common in the Mediterranean fringes. However, it is so much earlier there than in Britain that it is highly unlikely there is any meaningful connection between the two practices. Probably of more relevance are two fourth-millennium BC primary burials from northern Europe: Felsstalle, Baden-Württemberg (Orschiedt 2012, 217); and Chauveau CH1, Godinne-sur-Meuse (Toussaint and Becker 1994, 78–82). There was also a slightly later example from Resplandy Cave, Hérault, dating to the late fourth or early third millennium BC (Vander Linden 2006, 321). Overall, it seems that primary burial in caves was rare in this period throughout Europe.

Successive inhumation

The best-represented rite in the earlier Neolithic caves of Britain is successive inhumation. As discussed in Chapter 3, this is perhaps unsurprising; recent interpretations of chambered cairn burial deposits from the same period have suggested that it is also the commonest burial rite used in these monuments. Three cave sites which have already been discussed as part of Chapter 5 were almost certainly also places where successive inhumation took place: Thaw Head Cave, North Yorkshire; An Corran, Skye; and Raschoille, Argyll. However, there were at least five more Early Neolithic sites where this rite took place.

Bower Farm, Staffordshire (Appendix 1, number 7: NGR SK 0303 1954) is a small erosional rock shelter in a sandstone outcrop near Rudgeley. It was excavated by the Birmingham University Field

Archaeology Unit in 1979 following the discovery of human remains at the site. Cane and Cane (1986, 1–4) describe two female crania discovered at or close to the entrance to the rock shelter (see Figure 6.9). Their excavations also uncovered a relatively large assemblage of Mesolithic stone tools. The human bone from the site was reassessed by Blockley (2006) as part of her investigation into long-term trends in funerary behaviour. She established that there was a minimum of five individuals – three adults and two juveniles – from the site and that the assemblage included post-cranial elements. One of the adult individuals has an eighth-millennium BC radiocarbon date (Blockley 2006, 220). There is some doubt as to whether the dated sample was actually human bone (Meiklejohn et al. 2011, 34), but, if this date is accepted, it implies that the Neolithic burials at Bower Farm were of two adults and two juveniles and that all four were represented by both cranial and post-cranial elements.

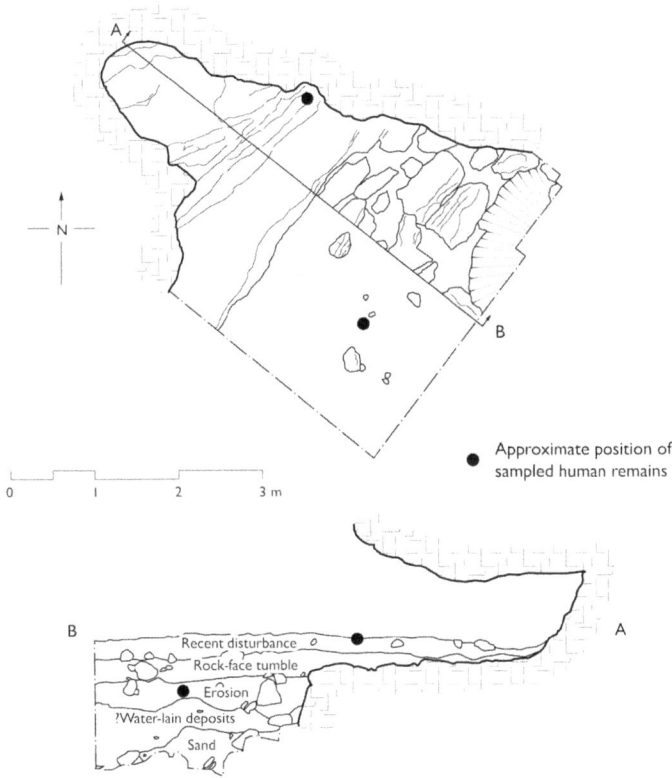

6.9 Plan and section of the excavated deposits within the Bower Farm rock shelter, showing the positions of the excavated human remains. After Cane and Cane (1986, Figure 3) with data from Blockley (2006, 202, 220, 395–397).

The two fourth-millennium BC radiocarbon dates come from a cranial fragment and a rib. They are highly similar and, if combined, could date a single burial event between 3600 and 3375 BC at two standard deviations. If the dates are modelled in OxCal 4.3 (Bronk Ramsey 2009; Reimer et al. 2013) on the assumption that they represent two events within a phase of burial activity, then, at two standard deviations, the earliest burial at the site took place between *3640 and 3420 BC* and the last burial between *3620 and 3400 BC*. The catalogued human bone listed by Blockley (2006, 395–397) includes phalanges, metacarpals, teeth and a navicular, along with cranial, axial and long-bone elements. Therefore, despite its highly fragmented condition, it seems likely to be the result of a successive inhumation burial rite. The recorded finds locations for the bone (see Figure 6.8) would tend to suggest that burial was taking place outside the rock shelter, with at least one of the bodies placed with their heads nearest the entrance.

Cave Ha 3, North Yorkshire (Appendix 1, number 14: NGR SD 7890 6624), is a medium-sized rock shelter, another of the cluster of sites in the south-west face of Giggleswick Scar (Figure 6.10). The site was excavated during the late 1940s and early 1950s, and a detailed archive report is held in Craven Museum (Leach 2006, 156–157). There was a deep tufa deposit within the shelter, which was still forming during the Early Neolithic (Pentecost et al. 1990, 95–96). It is therefore likely to have covered the bone as it was being deposited. The archive excavation notes show that bones probably were deliberately placed into the tufa (Leach 2006, 160). Some articulated bones were

6.10 The interior of Cave Ha 3, showing the area of the hearth and the niches in the rear wall of the shelter.

discovered towards the front of the shelter, associated with a large hearth, and others came from niches at the back of the shelter, where they were associated with two flint scrapers. The archive report refers to an adult foot set within tufa in one of the recesses at the rear of the shelter (Leach 2006, 157–158). This is presumably the articulated group of foot bones illustrated by Leach (2008, Figure 3.3).

There were four individuals from Cave Ha 3: a mature adult male and three juveniles (Leach 2006, 161–165). Leach (2006, 163) notes the presence of many labile elements within the bones ascribed to the mature adult male and interprets this as evidence that the body was intact and fleshed when it was buried in the tufa deposit. She also notes that one bone, the left tibia, had been split shortly after death in a similar way to cattle bones from the site. She interpreted this as evidence that the body was deliberately fragmented by people as part of the intermediary period, possibly also leading to the separation of the foot. Leach (2006, 160) also notes that very few of the bones showed signs of weathering, and there was only one bone with rodent gnawing. The tufa deposition seems to have acted to bury the bodies as they decomposed and protect them from both weathering and carnivore damage. The three juveniles from Cave Ha 3 were all very young. One was neonatal, another was between 9 and 12 months old and the third was approximately 2 years old (Leach 2006, 166–168). The preservation of skeletal elements in all three cases led Leach (2008, 47) to suggest that they were deposited as fleshed bodies in the niches at the back of the rock shelter.

Two skeletal elements from Cave Ha 3 were dated as part of Leach's (2006) research project. These were the splintered left tibia of the mature adult and the mandible of the 2-year-old child (Leach 2006, 169: and see Appendix 1). If these results are modelled in OxCal 4.3 (Bronk Ramsey 2009; Reimer et al. 2013), assuming that the two burials were events within a phase of funerary activity, then the first burial at the site probably took place between *3655* and *3520 BC* and the last burial between *3360* and *3040* BC, at two standard deviations.

It is clear from the detailed work carried out by Leach (2006) that the burial rite at Cave Ha was a form of successive inhumation. The cave environment, particularly the actively forming tufa, would have played a key role during the intermediary period by helping to preserve the bones from weathering and animal scavenging. Some tufa appears to be still actively forming in places on the rear wall of Cave Ha 3 today. However, it is also clear that there was active human involvement in fragmenting the body of the mature adult individual. The split tibia is rare evidence from the British Early Neolithic of the practice of manual disarticulation of the corpse during the intermediary period. As discussed in Chapter 2, this practice is much better attested in European Neolithic caves. For example, the early fourth-millennium BC human bone from Caverne B, Hastière, Belgium, has evidence of cut marks made by stone tools. The burial rite in this cave, and most other caves in the region, was successive inhumation (Cauwe 2004, 220). At Cave Ha 3, successive inhumation was a rite which brought people and cave processes together. Both would have acted to protect and fragment the remains in a proscribed manner over the intermediary period. There may have been a deliberate movement from the front of the shelter to the back as the intermediary period progressed. If this was the case, then we may also be able to see evidence for a simpler treatment for the very young children, who were probably placed directly into the niches at the back of the shelter.

The nearby site of Chapel Cave, Malham (Appendix 1, number 15: NGR SD 8810 6720), has been recently excavated and produced both Mesolithic and Neolithic human bone (Chamberlain 2014 and see Appendix 1). There were at least two people buried at the site in the Neolithic, and Christine Freeth's catalogue of the human bone from the site (Blockley 2006, 398–400) shows that a full range of skeletal elements was present. Therefore, it is probable that Chapel Cave was another site where successive inhumation took place.

Darfar Crag Cave, Staffordshire (Appendix 1, number 18: NGR SK 0975 5591), was excavated in 1986 without direct archaeological supervision, and consequently the contextual information from the site is somewhat limited (Blockley 2006, 208). The site, which is also known as Wetton Mill Fissure, is one of a group of three small caves in Darfar Crag. It has a small entrance, around 0.5 metres wide and similarly high, which leads to a small chamber around 5 metres deep. The two dates from this site were produced as part of a study by Blockley (2006) of long-term trends in funerary behaviour. Fortunately, this included a detailed examination of the osteological remains from Darfar Crag Cave, allowing some conclusions to be drawn about the possible funeral rites at this site. There were a minimum of five individuals – three adults and two juveniles – from the site (Blockley 2006, 213–214). The two Neolithic dates (see Appendix 1) come from one of the adults and one of the juveniles. These two dates are not sufficiently similar that it is likely that they represent a single event. If they are modelled on the basis that the burials represent a phase of activity, then, at two standard deviations, burial began between 3765 and 3640 BC and ended between 3630 and 3370 BC.

Blockley's (2006, 382–395) catalogue of the surviving remains included mandibles, teeth, bones of the arm and leg, many phalanges, carpal and metacarpal bones, vertebrae, patellae and sacrum. Where these elements can be ascribed to an individual, they show that all five of the bodies retained elements, such as the patellae, with labile articulations. Therefore, disarticulation seems to have taken place within the cave. The consistent presence of limb and trunk bones for all the individuals in the collection also suggests that the bodies were not moved after the intermediary period and that therefore the burial rite at Darfar Crag Cave was successive inhumation.

Hay Wood Cave, Somerset (Appendix 1, number 22: NGR ST 3398 5824), provides by far the best dating evidence for successive inhumation in the Early Neolithic, thanks to recent radiocarbon and dietary isotope research on the human bone from the site (Schulting et al. 2013). The site is a small, north-facing rock shelter which leads to an extremely narrow passage. Excavation work at the site was carried out between 1957 and 1971 by the Axbridge Caving Group and Archaeological Society (Everton and Everton 1972, 5). A 3.3 × 6 metre area of the interior of the shelter and the platform outside was excavated to a depth of up to 5 metres. The site produced a substantial assemblage of Romano-British and Iron Age pottery as well as an assemblage of Mesolithic worked stone, but no diagnostically Neolithic material culture (Schulting et al. 2013, 22). The deposits in the cave were considerably disturbed by badger burrowing, but the overall sequence can be seen in Figure 6.11. The uppermost layers, 1 and 2, were clay loams mixed with limestone fragments. The bulk of the Iron Age and Romano-British material came from these upper layers. Beneath this was what appears to have been a thick deposit of matrix-supported scree with many large angular limestone fragments, layer 3. This scree extended as far as the surface of the limestone bedrock in the western and eastern parts of the excavation. However, in the centre of the trench, there was a vertical rift in the bedrock. This extended into the rock wall of the shelter to form the entrance to a circular tunnel approximately 2 metres in diameter. The rift was filled with a reddish sandy loam, layer 4, while the tunnel fill was much more clay-rich. Most of the human bone was discovered in layer 3, in a disarticulated and co-mingled state, with some human bone, including one of the crania, coming from the fill of the tunnel (Everton and Everton 1972, 9–11).

The work by Schulting and colleagues (2013, 12–15) has provided an up-to-date assessment of the human bone assemblage. There were at least ten individuals buried at Hay Wood Cave: eight adults, an adolescent and a child of around 6 years old. Where a sex can be identified for the adults, three were women and three were men. Based on this initial assessment, which identified a large number of

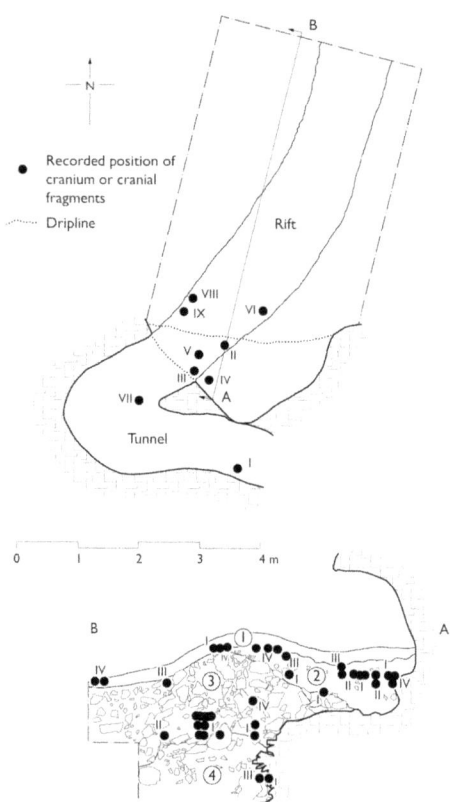

6.11 Plan and section of the excavated area at Hay Wood Cave showing the find locations for the human crania and cranial fragments (after Everton and Everton 1972, Figures 2, 3 and 4).

hand and foot bones within the assemblage, Schulting and colleagues (2013, 15) interpreted the burial rite as being successive inhumation. One cervical vertebra from the site was dated prior to the start of the recent research, and the radiocarbon programme produced seventeen new measurements from across all ten of the identified individuals (see Appendix 1). This means that we can be confident that the vast majority of the human bone from Hay Wood cave was deposited in the fourth millennium BC.

These dates included five sets of duplicate measurements on bones which were known to belong to the same individual. Each of these duplicate measurements were combined, and the resulting dates were modelled by Schulting and colleagues (2013, 17) on the assumption that the burial activity was a single phase of unknown duration. They were able to suggest, at two standard deviations, that burial at

Hay Wood Cave began between *3930* and *3715 BC* and lasted until *3580–3350 BC*. This model includes dates for two bodies that appear to be slightly earlier than the rest of the burials: the adolescent and the cranium II individual. Schulting and colleagues (2013, 17) considered the possibility that there were two successive phases of burial at Hay Wood Cave but concluded that the data was best explained by a single phase of longer duration. An alternative, as suggested by Weiss-Krejci (2012, 130), for the Iberian Copper Age caves mentioned here is that the burial activity took place towards the more recent part of the modelled range but that it included some curated bone. In this case, the burial rites at Hay Wood Cave would have been slightly more varied than the initial skeletal assessment suggested.

On balance, it is likely that bodies at Hay Wood Cave were placed in the tunnel and upper layers of the rift and left there over the intermediary period as they became disarticulated. The context descriptions provided in Everton and Everton (1972, 9–10) suggest that layer 3 formed rapidly, with large fragments of limestone eroding rapidly from the rock-shelter roof. The bodies were clearly accessible to both people and animals as they decomposed: Schulting and colleagues (2013, 13) note the presence of rodent tooth scores on human bone in their initial assessment. The fragmentation and disarticulation of the bones was caused as new bodies were added to the cave over the relatively short period that it was in use.

Successive inhumation burial was the funerary rite which drew most extensively on the active nature of both caves and environmental agents. Because the bodies did not physically move during the intermediary period, then the material narrative of changes is often easier to reconstruct. The interaction of multiple bodies and active cave processes, such as the tufa deposition noted here, would have allowed the ongoing temporality of the intermediary period to be understood. In these cases, the evidence of past burials and cave processes would have formed the circulating reference which linked one aspect of the burial practice to the wider narrative of the funerary rite. We can also see clear evidence of the continued input of living people into this narrative. At Cave Ha 3, some manual dismemberment of the body took place alongside the process of bodily decomposition. At Hay Wood Cave, a small amount of curated bone may have been added to the assemblage. At the earlier site of Thaw Head Cave, considered in Chapter 5, there is evidence that people opened and closed the dry-stone blocking of the cave at different stages in the intermediary period. Successive inhumation was also common in European caves in the fourth millennium BC. For example, it occurs in southern France at Les Grottes des Barbilloux, Lot-et-Garonne and L'Abri du Pas-Estret, Dordogne (Beyneix 2012, 225–226); at many sites on the Iberian Peninsula (Weiss-Krejci 2012, 129); and at Vogelherd, Barden-Württemberg (Conard et al. 2004, 200); and it was extremely common in both the Middle and Late Neolithic in Belgium (Cauwe 2004, 219–220).

Multi-stage burial

There are many caves for which, for one reason or another, it is not possible to suggest which kind of burial rite was in use. I have summarised the relevant details and available dating evidence for these sites in this section. The osteologically trained reader may well feel that many more of the sites I have discussed in the earlier parts of this chapter also belong here. However, in the examples used, I felt that some useful clues could point towards a likely interpretation. In this section, the most that can be said is that the combination of the passage of time and the processes of bodily decomposition and cave sedimentation created some kind of multi-stage burial for the fragmentary human remains that now survive.

The Neolithic human bone from Cathole Cave, Gower (Appendix 1, number 12: NGR SS 5377 9002), was discovered during excavations by a Colonel Wood around 1864. Analysis of contemporary reports and the archives from subsequent excavations by Charles McBurney (1958 and 1959) and John Campbell (in 1968) shows that some of the human remains came from the upper layers of the entrance fill. The human bone consisted of two crania and some other skeletal elements (Walker et al. 2014, 132–133). One of these crania was dated by Rick Schulting (OxA-16605: 4675 +/- 39 BP: Appendix 1), and would calibrate to between 3630 and 3365 BC at two standard deviations.

The Cathole date is potentially interesting because it overlaps with those from the Parc le Breos Cwm chambered tomb (Schulting 2007, 592). This site is visible from the cave, on the floor of the valley below. Schulting's (2007, 592–593) study of the stable isotope values from Parc le Breos Cwm and contemporary local cave sites, including Cathole, suggested that there was a slight dietary difference between individuals buried in the caves and those buried at Parc le Breos Cwm. If this interpretation of two separate burial populations is accepted, then it is unlikely that Cathole and Parc le Breos Cwm functioned together as part of the same set of funerary practices. Despite their proximity, bone does not seem to have been moved from one site to another. The extremely vague archaeological information we have about the original location of the Cathole human bone makes it difficult to interpret the specific burial rite at this cave. Interestingly,

Whittle and Wysocki (1998, 157–158) interpreted the Parc le Breos Cwm human bone as the product of two different burial rites. Bones from the chambers showed the high degree of weathering and carnivore modification typical of secondary burials, whereas the bones from the passage seemed to have been successive inhumations.

Little Hoyle Cave, Pembrokeshire (Appendix 1, number 28: NGR SS 1118 9997), is a small maze cave with both a vertical and horizontal entrance. The site has been investigated several times, most recently by a National Museum of Wales team between 1984 and 1990. There are both Late Upper Palaeolithic and Mesolithic lithics from the site. The cave itself is within Longbury Bank, an important early medieval settlement. There are reported to be up to eighteen individuals deposited beneath the central chimney feature (Green 1986, 101). There are four published dates from Little Hoyle Cave (see Appendix 1) but they have large errors, and there is a high degree of overlap between them. Therefore, it is not possible to be certain whether this represents a single phase of burial or a more protracted use of the cave. It is statistically possible that these dates could all result from a single burial event. If this assumption was true, then a combined date for this event calculated in OxCal 4.3 (Bronk Ramsey 2009; Reimer et al. 2013) would lie between 3605 and 3380 BC. If the four dates were to be modelled on the alternative assumption that burial was a longer-term process, then burial began between *3795* and *3550 BC* and the last dated burial took place between *3645* and *3445 BC*. Burial at Little Hoyle certainly began during the Early Neolithic and was over considerably before 3000 BC.

Burial practices are hard to reconstruct in detail for Little Hoyle, as the dated bone comes from the early excavations. It is tempting to regard the eighteen reported bodies as an example of successive inhumation. However, given the lack of detailed stratigraphic information and up-to-date skeletal analysis, it is probably preferable to interpret the site conservatively as an example of a multi-stage burial, without attempting to specify the particular rite which was used.

There is a single radiocarbon date on a molar from Cattedown Cave, Devon (Appendix 1: OxA-15256: 4990 +/- 32). This is one of the few surviving elements from a much larger collection of human bone which was recovered from this cave under salvage conditions in 1887 (Chamberlain and Ray 1994, 30–31) and which was subsequently badly damaged by the bombing of Plymouth in 1941. The original account (Worth 1887, 110) of the discovery of the human remains makes it clear that these bodies were at least partially articulated when discovered, although it should be borne in mind that part of the excavation was carried out by blasting. The human bone came from a breccia deposit partially covered by a stalagmitic floor within the northern chamber, and it was claimed that they were associated with extinct mammalian remains. Worth (1887, 111) gives a minimum number of individuals of fifteen for the whole cave, but it is unclear what criteria he used to arrive at this figure. He also stated that the assemblage included examples of 'Every bone of the human frame' (Worth 1887, 112). The relative completeness of the bone assemblage suggests that this is an example of deliberate multi-stage burial in the depths of what was formerly an extensive system (Chamberlain and Ray 1994, 30). However, in view of the salvage nature of the excavations and the history of the archive since, a more detailed interpretation is not possible.

Markland Grips, Derbyshire (Appendix 1, number 30: NGR SK 510 751), was one of three cave sites in this valley excavated by A.L. Armstrong in 1924. The human bone at this site was behind two separate drystone walls at the back of the cave. There were a minimum of five individuals from the site. Archive records suggest that the number of surviving bones was small and that they were disarticulated when found (Hedges et al. 1996, 399–400). The Markland Grips remains seem to have been directly associated with sherds of four Early Neolithic bowls (Griffiths 2011, 86).

There are two very similar published dates from Markland Grips Cave (see Appendix 1). Analysis by Griffiths (2011, 85–86) shows that it is statistically possible that the dated individuals could have died at the same time and that this may be a single burial event. On this basis, a combined date for this event calculated in OxCal 4.3 (Bronk Ramsey 2009; Reimer et al. 2013) would lie between *3585* and *3365* BC. Alternatively, burial at the site may have been a longer-term process. Modelling the dates on this assumption would suggest that burial began between *3710* and *3445* BC and the last dated burial took place between *3650* and *3425* BC. Frustratingly, the extant information about the Markland Grips Cave makes it a difficult site to interpret satisfactorily. As Griffiths (2011, 86) points out, the use of dry-stone walling to enclose the human remains may echo the construction practices at chambered cairn sites. Dry-stone walling is also known at other cave sites: Thaw Head Cave in North Yorkshire was discussed in Chapter 5; it also occurs at Middle and Late Neolithic cave sites in Belgium (Cauwe 2004, 220); and Dowd (2015, 113) notes evidence for the blocking of a number of Neolithic burial caves in Ireland. However, in view of the lack of detailed contextual and osteological information for this site, the funerary rite which led to the 'sparse' human remains behind these two walls cannot be more precisely interpreted than as a multi-stage burial.

The platform outside Picken's Hole, Somerset (Appendix 1, number 38: NGR ST 3969 5500), was excavated between 1961 and 1967 (ApSimon 1986, 55; Tratman 1964, 1–2). There was a series of Pleistocene deposits beneath the modern topsoil (see Figure 6.12). Layer 3 was a silty matrix-supported scree which contained 53 pieces of worked stone and fragmentary animal bones, including spotted hyaena, woolly rhinoceros, horse, red deer, reindeer and large bovid species. There were also two human teeth from this layer. Radiocarbon dates on the animal bones were entirely consistent with the interpretation of the site as an Early Upper Palaeolithic occupation site (ApSimon 1986, 56). However, one of the teeth was radiocarbon-dated (OxA-5865: see Appendix 1) and gave a result which would calibrate to between 3695 and 3380 BC at two standard deviations. There is some evidence for disturbance of layer 3, so it seems likely that the human remains at Picken's Hole were originally deposited on the top of the Late Pleistocene scree outside the cave. Other human teeth were reportedly recovered from unstratified deposits at the site (Hedges et al. 1997, 446). In view of the very small number of surviving teeth, it may be that Picken's Hole was another site which was used during an intermediary period before the bulk of the remains were moved to another site for secondary burial. However, a small number of successive interments, or even a primary burial, on the scree surface would have also left few fragmentary remains like this if they were exposed

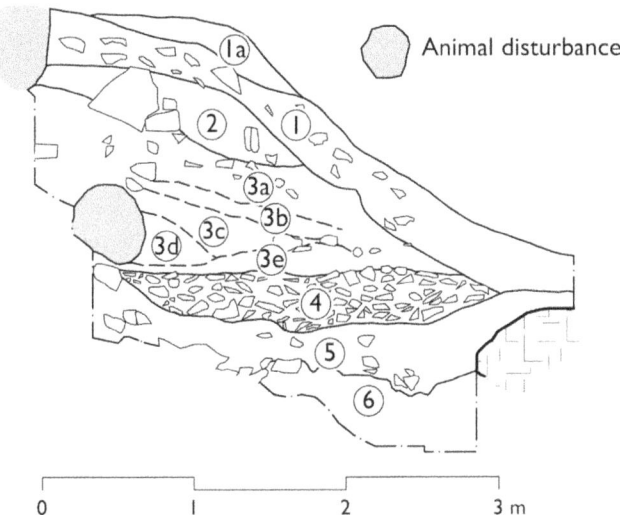

6.12 Section through the deposits outside Picken's Hole, showing the layer of matrix-supported scree (layer 3) where the human teeth were discovered (after Tratman 1964, Figure 18).

to the actions of the elements and animals for long enough. Therefore, it seems best to interpret Picken's Hole as a multi-stage burial.

There are also seven cave or rock-shelter sites in Britain with broadly earlier Neolithic radiocarbon dates but where there is not yet satisfactory evidence to interpret the type of funerary rite at even a very general level. However, these sites do potentially give us useful information about the kind of locations where human remains were being deposited in the period. I will return to this point in Chapter 8. The sites, which are listed in Appendix 1, are Ash Tree Shelter, Derbyshire; Foxhole Cave, Paviland, Gower; Kent's Cavern, Devon; King Arthur's Cave, Herefordshire; Ogof y Benglog, Caldey; Ossum's Crag Cave, Staffordshire; and Tornewton Cave, Devon.

Conclusions

The cave burials of the earlier Neolithic in Britain appear to have been relatively varied. Despite the apparent preponderance of successive inhumation, there are a significant number of sites with good evidence for different practices. This suggests that the diversity of rites observable right at the beginning of the fourth millennium BC continued into the Early Neolithic. During the first half of the Neolithic, it is probably more accurate to talk about cave burial *practices* rather than a cave burial *rite*. Many of these practices were not exclusive to caves. Successive inhumation was clearly a rite which was appropriate for both chambered cairns and caves. It may well have occurred at other subterranean locations, such as flint mines. There has been some discussion as to whether caves and monuments used in this way were perceived in the past as equivalent spaces. Barnett and Edmonds (2002, 119) thought that they probably were, sharing a common set of properties. Dowd (2015, 110), by contrast, thought that, for the Irish Neolithic at least, there was a genuine distinction between what happened in caves and what happened in monuments. Her view may be supported by the fact that, while the two kinds of burial site share the practice of successive inhumation, the other burial rites considered in this chapter do not seem to occur at chambered cairns. None of them however, appear to be exclusive to caves. There is evidence for the putative 'cult of the head' identified here at causewayed enclosure sites such as Etton, Cambridgeshire (Pryor 1998, 271), and possibly from Staines (Schulting and Wysocki 2005, 128). The burials in midden material discussed in Chapter 5 show a similar connection between both cave and open-air middens. Likewise, Early Neolithic primary burial was rare, but it seems to have occurred both in caves and limestone fissures and in flat graves. Apart from the case of primary burial, the important linking factor between the funerary rites discussed in this chapter is that, in all of them, living people were actively involved. In Early Neolithic cave burials, the intermediary period was a time when people would have not only observed the material narrative of change, being able to read the clues which told them which parts of the funerary rite were appropriate at which time, but also actively intervened to ensure they happened. Caves and landscapes would have acted as circulating references, linking rites to particular times and places, and bodies would have provided an entangled directionality to the rites, but much of the practice was carried out by living people. As I will show in the next chapter, this emphasis seems to have changed in the Middle and Later Neolithic.

7

Deep time

Introduction

The sites reviewed in the last chapter demonstrated two important points. The first was that Early Neolithic cave burial was a relatively diverse set of practices, often connected to other kinds of places. Although caves and rock shelters provided one kind of active environment and helped to constitute the temporality of these rites, there is evidence that the rites could equally well have taken place in other kinds of location. The second was that most Neolithic human remains in caves date to the early part of the period. The Early Neolithic bias in dates for human remains in caves has been noted previously (Chamberlain 1996, 950; Schulting 2007, 586). However, many of these sites have only a single radiocarbon date. Sites with multiple dates on human bone often also produced evidence for activity later in the period. For example, An Corran Rock Shelter, Skye, which was considered in Chapter 5 because it was an example of a midden burial site, has evidence for successive inhumation as late as the beginning of the Early Bronze Age. In general, cave burial is largely an early phenomenon, but we should be wary of assuming that sites with a single Early Neolithic date were only used in that period. By the end of the fourth millennium BC, there appears to have been a number of earlier burial sites which were still being used alongside a smaller number of sites which first began to be used from the Middle Neolithic onwards. The locations of all the sites discussed in this chapter are shown in Figure 7.1.

Successive inhumation

The successive inhumation rite which was so common in the Early Neolithic was apparently still being practiced into the Middle Neolithic. Some of the sites discussed in this section began to be

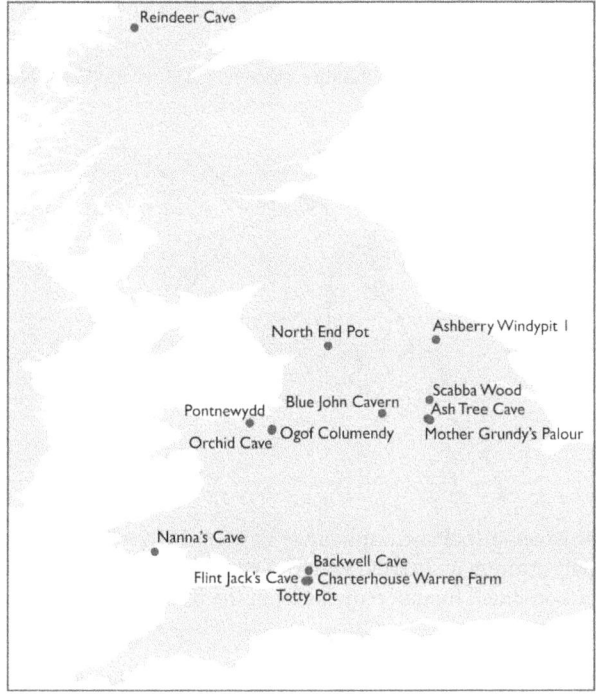

7.1 Location map for the sites discussed in Chapter 7. The base mapping includes data licenced from © EuroGeographics.

used at the end of the Early Neolithic, but they have been discussed in detail here as most of their use seems to have fallen into the Middle and Later Neolithic. Even more than seems to have been the case in the Early Neolithic, these were sites where the primary agents during the intermediary period were bodily decomposition and cave processes. Successive inhumation took place in at least five caves during the Middle Neolithic.

Reindeer Cave is one of a group of four caves in a limestone crag above Allt nan Uamh, Inchnadamph (Appendix 1, number 41: NGR NC 2682 1704), just south of Loch Assynt in Sutherland. The dated human bone comes from excavations carried out by James Cree in 1926 (Callander et al. 1927), which were subsequently re-interpreted and published by Lawson (1981) and Saville (2005). Human bone was found in two places within the first chamber of Reindeer Cave (see Figure 7.2). Both finds were within the uppermost, red-clay, layer of the cave fill, and all the human remains are likely to be Holocene. A bone pin fragment and a cranium without either maxilla or mandible were found within a small cist-like structure made from two limestone slabs. A sacrum and several vertebrae were also found within

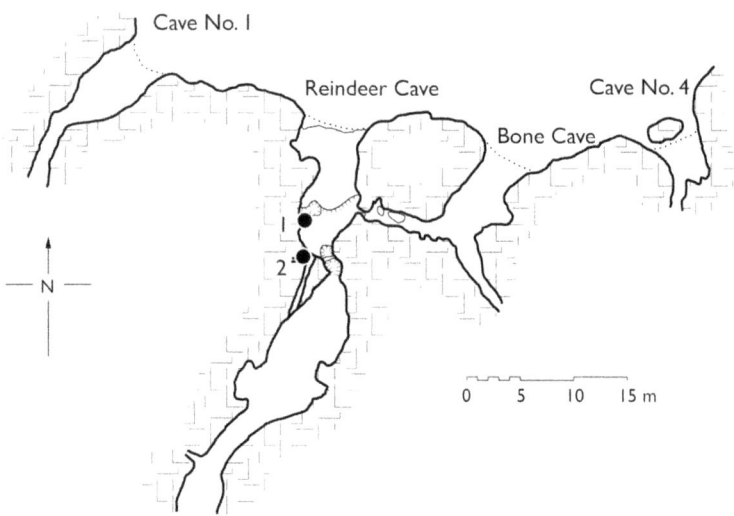

7.2 Plan of the caves at Inchnadamph (after Lawson 1981) showing the location of (1) the human cranium associated with disarticulated vertebra and sacrum and (2) the radiocarbon-dated human remains from the fissure at the back of the first chamber.

this structure (Lawson 1981, 14–15). The pin is walrus ivory and has been radiocarbon-dated to the early medieval period (Saville 2005, 352), which, however, may not tell us anything useful about the date of the human remains.

A burial which was described by the original excavators as a substantially complete but disarticulated juvenile skeleton was discovered in a fissure at the back of the first chamber (Lawson 1981, 15). Four radiocarbon dates were obtained on bones from this deposit in 1995 (Hedges et al. 1998, 438). These suggest that it dates to the mid to late fourth millennium BC (see Appendix 1). However, Saville (2005, 356–358) has shown that this deposit includes bones from at least three individuals – two juveniles and one adult – and so the original interpretation of this as a single burial should be revised.

Saville (2005, 358) noted that the three later dates could be combined and may relate to a single burial event; indeed, they may possibly all belong to the same juvenile individual. If all the dates are modelled in OxCal 4.3 (Bronk Ramsey 2009; Reimer et al. 2013) on the assumption that they represent a single phase of activity, then this suggests, at two standard deviations, that burial began between 3625 and 3365 BC and that the last burial was placed between 3335 and 3025 BC. The adult metatarsal dated by OxA-5761 is significantly earlier than the results on the juvenile material. However, even if the dates are modelled assuming that this burial is the first in a sequence, the overall probable duration of use for the cave is not significantly altered.

Although the human bone from the site could not be located prior to Saville's (2005, 357) work, an archive catalogue of the remains survives from 1962. This shows that the range of skeletal elements preserved at Reindeer Cave included both cranial and post-cranial material. The surviving material was dominated by bones from the limbs and trunk but also included two metatarsals, one of which had rodent tooth scoring. The excavation notebooks (Saville 2005, 348–349) described the bones as being within the fill of the large fissure at the back of the first cave. This fissure slopes steeply downwards to connect to the lower chamber to the south (see Figure 7.2). Cree and Callander were of the opinion that a skeleton had been placed head down into the fissure and that it had subsequently been disarticulated through a combination of gravity, moving cave sediment and animal burrowing. I think that we can accept this explanation for the disarticulated state of the bone. It is also likely that smaller elements have been moved a considerable distance further into the cave. The radiocarbon evidence and the skeletal catalogue show that this was in fact a repeated process, involving at least three different bodies in a multi-stage rite. The surviving presence of metatarsals may indicate that this was an example of successive inhumation, with the bodies having been placed at the back of the first chamber and gradually moved and disarticulated by the cave processes noted here. However, in view of the absence of systematic data on the weathering of the bone and the fact that only two metatarsals seem to have survived from the hands and feet of three individuals, it is also possible that the back of Reindeer Cave was used as a place of secondary burial.

Backwell Cave, Somerset (Appendix 1, number 5: NGR ST 4924 6801), is a small cave, around 2.7 metres deep by 1.2 metres wide, which was discovered and partially emptied in 1936. The clearing of the cave led to the discovery of human and animal bones, and in 1937 an excavation of the site was led by F.K. Tratman (1938). Despite the relatively early date of excavation, like Chelm's Combe and Hay Wood Cave, the Backwell Cave archive provides a good level of both archaeological and osteological detail. Tratman's (1938, 58–61 and see Figure 7.3) report shows that the sequence in the cave and the small platform beyond it was relatively simple. At the time of its original discovery, the cave was filled, almost to the roof, with what appears to have been a matrix-supported limestone scree. This deposit extended as a talus slope over the platform and was up to 1.5 metres deep. At this depth, there was a layer of calcite deposition which covered parts of the scree. Beneath the calcite was a further

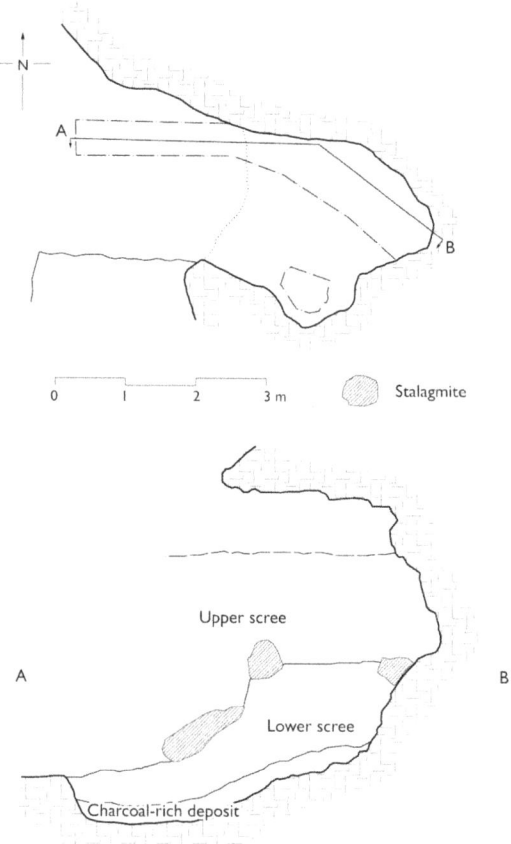

7.3 Plan and section of the excavated deposits in Backwell Cave. After Tratman (1938, Figure 23).

scree deposit which overlay, over the whole area of the cave floor, a charcoal-rich layer around 0.25 metres deep. This layer contained the human skeletal material, a smaller number of animal bones, some sherds of Romano-British pottery, two worked pieces of bone and two flint artefacts: a leaf-shaped arrowhead and a knife (Tratman 1938, 62–64). The cave deposits were somewhat disturbed by badger burrowing, and some of these finds were recovered from the spoil heap from the 1936 clearing of the cave rather than from Tratman's excavations, so this layer does not represent a sealed context of a single date (Tratman 1938, 67). Except for the lithics, which Tratman (1938, 63) considered to be residual, the artefacts are likely to belong to the first two centuries AD. Tratman (1938, 65–66) worked on the understandable assumption that the human and animal remains were of a similar date and that the whole cave was used for burial in the Romano-British period.

Two vertebrae from the collection were dated by Alison Roberts (Appendix 1: BM-3099: 4510+/-40 BP); this result would calibrate to between 3360 and 3090 BC at two standard deviations, showing that at least some of the human remains from the site dated to the Middle Neolithic. The problem here is that, unlike at Hay Wood Cave, where the recent dating programme has provided clear evidence that all the identifiable individuals were part of a single phase of burial activity (Schulting et al. 2013), we do not know how many of the Backwell Cave burials are Neolithic.

In Tratman's (1938, 71–74) report, the human bone was catalogued and studied by Prof. E. Fawcett. He established a minimum number of individuals for the site of fifteen adults and three children between 6 and 8 years old. This figure was based entirely on mandible fragments, and a modern re-assessment would probably produce a different result. Tratman treated the human bone as the result of a single phase of burial. He specifically attempted to interpret the kind of funerary rite which had taken place and concluded that it was successive inhumation (Tratman 1938, 65–66). He based this interpretation on the plentiful presence of hand and foot bones, along with all the major skeletal elements. He also suggested there was a noticeable preponderance of left-sided elements in the assemblage. He interpreted this as the result of differential preservation in a rite where bodies were being consistently laid on their left sides. Where this supposed bias can be checked in the published catalogue, it is not particularly overwhelming. For example, only 61% of the recorded mandible fragments are left-sided (Tratman 1938, 71–73). Assuming that Tratman was right to treat the human bone as a coherent assemblage and that it was all Middle Neolithic in date, it seems likely that the bodies were introduced into the cave as successive inhumations. There is no published data on carnivore damage to the bone or on the amount of weathering, so it is not clear if they were protected in any way during the intermediary period. The degree of fragmentation and disarticulation in all the bodies may suggest that they were not.

Nanna's Cave, Pembrokeshire (Appendix 1, number 32: NGR SS 1458 9698), is a deep rock shelter or shallow cave in cliffs on the north coast of Caldey Island. An important assemblage of Late Upper Palaeolithic and Mesolithic stone tools came from the site, along with Neolithic, Bronze Age, Iron Age and Romano-British artefacts. There have been excavations at the site at various dates since 1911 (Davies 1989a, 84). The overall sequence of deposits in the cave was established in excavations carried out by James Van Nédervelde in

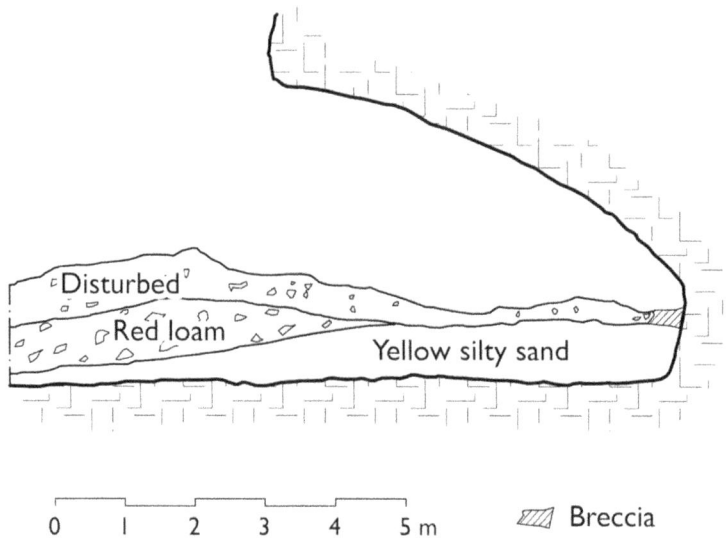

7.4 Section through the deposits in Nanna's Cave (after Lacaille and Grimes 1956, Figure 13).

1950 and 1951 (Lacaille and Grimes 1956, 99–103 and Figure 7.4). Within the cave, and covering the platform outside, was a layer of mixed limestone, cave earth and 'stalagmite' fragments, which was the disturbed spoil from earlier excavations within the cave. This layer, which was up to 0.6 metres thick, contained finds of a wide variety of different dates. At the back of the rock shelter, some calcite was found still *in situ* at this level. Given the open nature of the cave site, the calcite deposit is highly unlikely to have been stalagmite. It is probable that it was the cemented breccia described by Davies (1989a, 84) as being at the back of the cave. The presence of this layer is important because human bone recovered in earlier excavations was described as 'cemented together by stalagmite' (Lacaille and Grimes 1956, 97). Beneath this layer there was a 1-metre-deep layer of reddish loam colluvium with limestone fragments. This in turn overlay a yellow silty sand, described by Davies (1989a, 84) as a raised beach platform, which extended to the back of the cave. The most recent work was carried out by Davies and Van Nédervelde (1976) between 1973 and 1986. They recovered further human bone and established that there were the remains of a minimum of four people in the cave: three adults and one juvenile (Davies 1989a, 84). The human bone was probably deposited on the surface of the raised beach inside the cave, where some of it was encrusted in the cemented breccia deposit.

Two pieces of the human bone from Nanna's Cave were dated by Schulting and Richards (2002a, 1014), and a further six were sampled for dietary isotopes. The dates (see Appendix 1) were both Middle Neolithic and could conceivably date a single event, in which case a combined date would calibrate to between 3295 and 3095 BC at two standard deviations. If we assume that the burials were part of a phase of activity, then this began between *3495* and *3170 BC* and was over by *3295–3150* BC. The stable isotope values for all eight of the sampled bones were extremely consistent, which tends to suggest that all the human bone was part of the same group of burials and that it all dates to the Middle Neolithic.

The burial rite at Nanna's Cave can only really be reconstructed tentatively. The encrustation of some of the bones suggests that bodies were placed towards the back of the cave, near the surviving cemented breccia deposit. The bones sampled by Schulting and Richards (2002a, 1040) included two patellae and a phalanx, which is suggestive of a successive inhumation rite. It is also notable, as was the case at Reindeer Cave, that the number of individuals buried at the cave was relatively low.

The evidence from another Middle Neolithic site, Scabba Wood Rock Shelter in South Yorkshire (Appendix 1, number 43: NGR SE 5269 0196), is much better. The site has been excavated twice. Human bone was discovered in the rock shelter by the landowner in 1991, and as a result, an evaluation was carried out by South Yorkshire Archaeology Service (Chadwick 1992). This evaluation was followed by a research excavation in 1998 carried out by the University of Sheffield (Buckland et al. 1998). Scabba Wood Rock Shelter is a low, shallow overhang beneath a limestone outcrop. The shelter is nowhere more than 1 metre deep, and before excavation, there was only a small gap between the top of the fill and the shelter roof. The excavations covered a 7 × 10-metre area within and to the west of the rock shelter itself (Buckland et al. 1998, 6–9 and see Figure 7.5). Beneath the topsoil over the entire excavated area were layers of humic soil with many limestone fragments. These layers were clearly open in texture and actively reworked by burrowing animals. There were post-medieval and Roman finds in this layer along with eighteen flakes and blades of worked stone, which were probably Early Neolithic. Beneath these layers was a more compact orange-brown loam with many limestone fragments. This, too, had finds of a wide range of dates within it, including both Roman and Iron Age pottery and parts of five leaf-shaped arrowheads. There were some human remains from this layer reported by Buckland and colleagues

(1998, 7), and the spread of human material recovered by Chadwick (1992, 5) came from the lower interface between this layer and a clay layer beneath. There was a second area excavated to the north of the main shelter (Buckland et al. 1998, 9); a single fragment of adult cranium came from the upper layers here, but excavation was not continued in this area.

The human bone from both excavations was reassessed by Buckland and colleagues (1998, 9–11). They concluded that there were at least three, and possibly as many as seven, people buried at the site. Individual A was an adult man, between 20 and 30 years old, and was the best-preserved body. Approximately 25% of the skeleton of this individual survived (Rega in Chadwick 1992, 12).

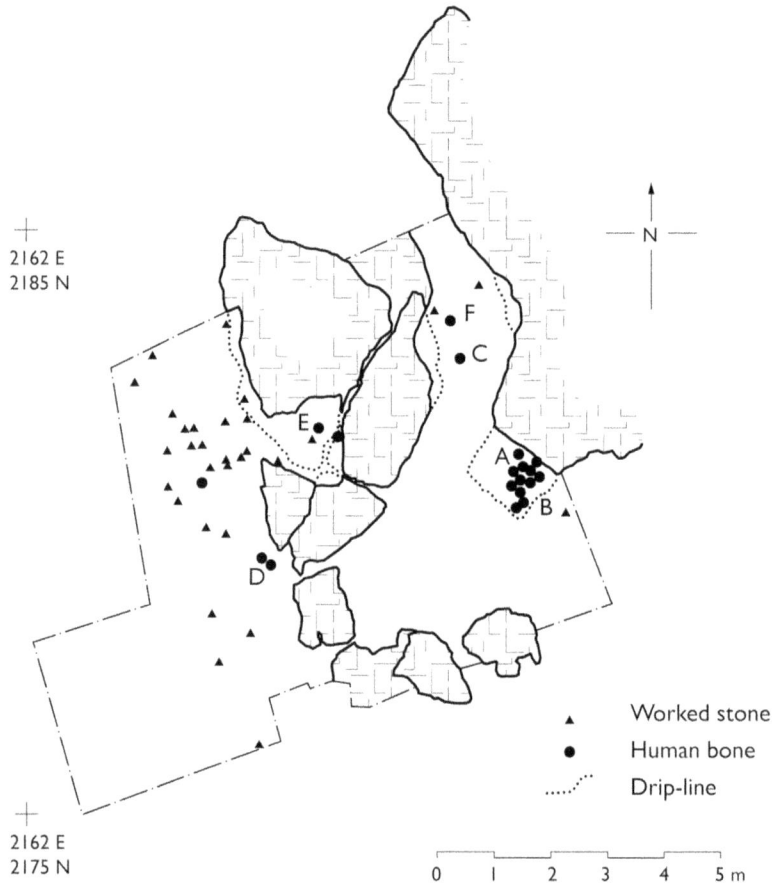

7.5 Plan of excavated area P at Scabba Wood Rock Shelter, showing the location of the human remains and Neolithic worked stone (after Chadwick 1992, Figure 4; and Buckland and colleagues 1998, Figures 6, 8, 10 and 11).

Individual B was a juvenile of between 12 and 15 years old and was only represented by disarticulated teeth from both the 1992 and 1998 excavations. Individual C was represented only by tooth and mandible fragments and a single manual phalanx and was an adult of over 40 years old. Two teeth show that individual D was a child of between 6 and 9 years old. There were a further three individuals identified from tooth and bone fragments, which may however, actually be widely dispersed parts of individuals A and B (Buckland et al. 1998, 9–11). The dated bone comes from the scatter of remains recovered in 1992 (Buckland et al. 1998, 11). It is therefore almost certainly part of individual A. This result (Appendix 1: UB-3629: 4590 +/- 30 BP) would calibrate to between 3500 and 3125 BC at two standard deviations.

The burial rite at Scabba Wood can be reconstructed with a reasonable degree of confidence. The adult man, individual A, was probably the last body to be placed at the site. The catalogue of bones provided by Rega (in Chadwick 1992, 12) includes mandible fragments, parts of all the vertebral column from the cervical vertebrae to the sacrum, ribs, parts of the pelvic girdle and bones from both the left and right arms and legs. Therefore, all the major parts of the body seem to have been brought to the site, however fragmented they subsequently became. There were also 21 hand and foot bones which probably belonged to this man, excellent evidence that the body was still at least partly fleshed when it was placed in the rock shelter. Rega (in Chadwick 1992, 12) noted that the bone fragments were highly weathered, but she did not find any evidence for animal scavenging. Buckland and colleagues (1998, 9–10) noted the predominance of teeth and phalanges in the whole assemblage and suggested that these, denser, elements were differentially better preserved. I would suggest that Scabba Wood was being used for successive inhumation from the later part of the Early Neolithic, with bodies being placed in turn in the narrow space between the rock shelter and the limestone blocks to the west. Once there, they may have been protected from animal scavenging by temporary barriers such as fences or hurdles. After an extended intermediary period, these barriers were removed and the bone became further disarticulated. The date of death for individual A given by the radiocarbon result suggests that the most recent of these burials took place in the Middle Neolithic. Other kinds of Neolithic deposition may also have taken place at the site. The worked stone (see Figure 7.5) seems to have been placed largely outside the area used for funerary rites. It is possible these artefacts were deposited during intermediary periods, when the rock shelter itself was shut off to human access.

A much longer sequence of burial, which included Middle Neolithic activity, is evident at Totty Pot, Somerset (Appendix 1, number 48: NGR ST 4825 5357), and in this case, the human bone comes from approximately 10 metres into an underground system. Totty Pot is a vertical fissure which leads, after a short squeeze, into a multi-chambered cave (Gardiner 2016, 42–43 and see Figure 7.6). The site was discovered in 1960 and was explored by the Wessex Caving Club until at least 1965. Initial digging at the site was purely focussed on opening the cave for underground exploration. Following the recognition of human and animal bone and prehistoric worked stone, the team began to keep notes on the location of finds and, in 1965, established a recording grid within the cave itself. Gardiner (2016, 43–52) has reviewed the surviving archive information and reconstructed the former position of some of the finds. I have followed her interpretation of the stratigraphy. It seems clear that the archaeological materials come from layers of clast-supported scree and friable cave earth (Gardiner 2016, 47–48). These layers covered a deposit described as a 'tufa floor', and most of the human bone was described as coming from immediately above this floor in two areas (Gardiner 2016, 49 and see Figure 7.6). Some of the surviving human bone analysed by Schulting, Gardiner and colleagues (2010, 80) was partially coated in calcium carbonate from the top of this deposit. Finds from Totty Pot include a small assemblage of Late Mesolithic microliths, a barbed and tanged arrowhead, human and animal bone and Early to Middle Bronze Age pottery. A radiocarbon date on a human tibia from the southern part of the cave (Ambers and Bowman 2003, 532: BM-2973: 8180 +/- 70 BP) showed that some of the human bone was Mesolithic. The human remains from the site have had a problematic curation history, but a partial assemblage of around 60 elements, mostly from the 1963 season, was available for re-study by Schulting, Gardiner and colleagues (2010, 77). This probably represents about half of the material which was originally excavated.

This study established that the surviving material included the remains of at least seven people, including four adults (two women and two men), a child of around 10 years old and two younger children (Schulting, Gardiner et al. 2010, 78). The radiocarbon dating programme established that one of these men was Mesolithic, with a date (OxA-16457: 8245 +/- 45 BP) which is very similar to the other Mesolithic result from the site. The other five individuals dated, however, covered a range of time from the earlier Neolithic until the very end of the period (Appendix 1). Schulting, Gardiner and colleagues (2010, 81) considered that there were three separate episodes of burial

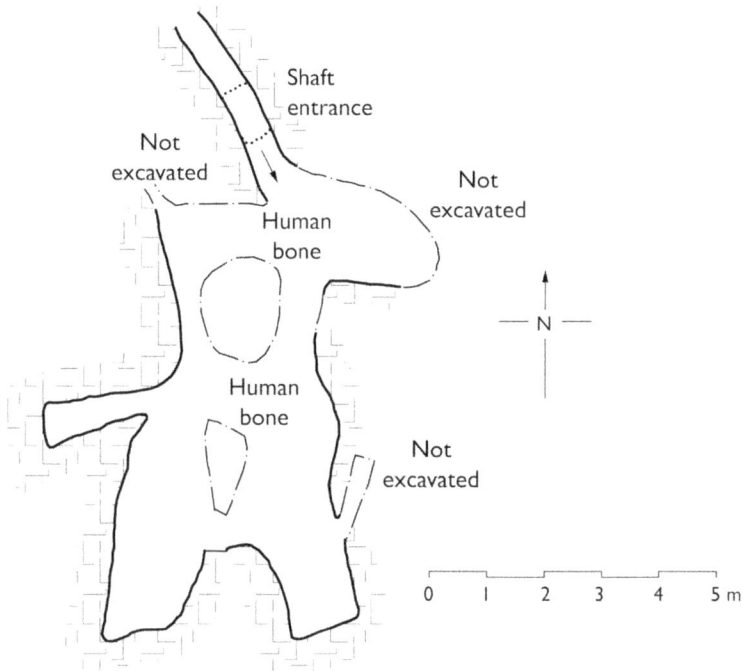

7.6 Plan of the interior of Totty Pot and the location of some of the archaeological material (after Gardiner 2016, Figure 13).

within the Neolithic at Totty Pot. On this assumption, the dates can be modelled in OxCal 4.3 (Bronk Ramsey 2009; Reimer et al. 2013), and, at two standard deviations, this suggests that the first Neolithic individual to be buried at the site died between *3630* and *3370 BC*. Following the same assumptions, the three Middle Neolithic burials took place between *3355* and *2930 BC*, and the final dated burial took place between *2840* and *2465 BC*.

As there was demonstrably Neolithic burial activity at Totty Pot for at least 550 years, and we know that the surviving bone assemblage represents only a portion of what was originally in the cave, it is perhaps a little optimistic to attempt to discuss *the* burial rite in this case. Over this long period, it is likely that there were considerable variations in funerary practice. However, several kinds of rite can be ruled out (Schulting, Gardiner et al. 2010, 78). None of the bone in the assemblage shows signs of sub-aerial weathering, there is little surface erosion of any kind and the only tooth marks are rodent gnawing on two bones. Therefore, none of the bodies were exposed outside the cave in the intermediary period. There were also no signs that the bodies had been manually defleshed; none of them were burnt or cut with stone tools. It seems likely that the dead entered the cave as complete fleshed bodies. Schulting, Gardiner and colleagues (2010, 87) thought that successive inhumation was the likely rite, supported by the presence of a small number of phalanges in the assemblage. If this was the case, were the bodies carried into the underground chambers by living mourners? Gardiner (2016, 66) considered the possibility of manually handling bodies down the shaft and into the chamber and concluded that it would have been 'difficult, but not impossible'. The Late Bronze Age example of Robber's Den Cave, Co Clare, where access to a burial site required climbing a rock face and several hours of caving (Dowd 2015, 145), shows that such a practice was possible. An alternative interpretation is suggested by the sketch sections from the original explorations (Gardiner 2016, Figure 5), which show how the surface of the deposit sloped away from the base of the entrance rift. The modern example of Jama-Bezdan, Hrgar (Simmons 2002), which was discussed in Chapter 3, shows that whole bodies which were dropped down a vertical shaft onto an unstable talus slope would become disarticulated and dispersed in a way which is entirely compatible with the reconstructed positions of the human bone.

There are also some Middle Neolithic dates from sites where the evidence is much more fragmentary. The single metatarsal from Pontnewydd, Denbighshire (Appendix 1, number 39: NGR SJ 0152 7102) comes from an individual who died between 3370 and 2930 BC (OxA-5820: 4495 +/- 70 BP: Aldhouse-Green et al. 1996, 446). The bone was found in the spoil-heap from nineteenth and early twentieth-century excavations at the site, and there are other earlier and later Holocene human remains from the same area (Aldhouse-Green et al. 1996, 445). Therefore, the most that can be confidently said about Pontnewydd is that some kind of funerary rite took place at or near the cave in the Middle Neolithic. At another North Welsh site, Ogof Columendy rock shelter, Flintshire (Appendix 1, number 35: NGR SJ 2020 6277), ongoing work by Clwyd-Powys Archaeological Trust has built on excavations by North Wales Caving Club in the 1970s (Davies 1989b, 99). A radiocarbon date of 4408 +/- 33 BP (SUERC-66486) is reported from a molar at this site (Ebbs 2017), which indicates a calibrated range of 3310–2915 BC. The evidence from Flint Jack's Cave, Somerset (Appendix 1, number 19: NGR ST 4632 5381) is slightly more complete. Ambers and Bowman (2003, 532) report that the human femur they dated from this cave was part of a collection of human bone including parts of four individuals. The cave itself is a small rock shelter which was excavated in the late nineteenth century. Oakley (1958, plate 7) reviewed the evidence for the original finds location of this bone and reproduced a nineteenth-century photograph which appears to show the bones *in situ* and partly cemented by tufa deposits towards the back of the rock shelter. The assemblage included at least two crania, but without further osteological details, it is difficult to suggest what kind of funerary rite led to this deposit. The date on the femur (BM-2839: 4430 +/- 80) shows that one of these individuals died between 3345 and 2915 BC.

In every place where we can reconstruct the kind of funerary rites in use in the Middle Neolithic, some form of successive inhumation seems to have been in use. The general impression is that the intermediary period was something which took place within caves, somewhat secluded from the activities of the living mourners. Even at Scabba Wood Rock Shelter, where burial could not be completely hidden, it seems to have been partly protected, perhaps by temporary screens. Alongside this, at sites like Reindeer Cave and Totty Pot, cave morphology and the movement of cave sediments seem to have been actively involved in the disarticulation of bodies. Therefore, the temporality of the rite of successive inhumation at this period was being constituted by the actions of caves and decomposing bodies, rather than the intervention of living people. This has interesting implications, which will be explored further in Chapter 8, for the overall duration of successive inhumation at each site in these later periods. The passage of time indicated by bodily decomposition and cave sedimentation also seems to have become something which was supposed to happen out of sight; cave burials in the Middle Neolithic were more likely to be found deeper into the system. Something similar may be noted in Belgium, where in the late fourth and early third millennia BC there was an emphasis on successive inhumation burial in the large number of Seine-Oise-Marne Late Neolithic burial caves. The practice of blocking or closing some of these caves with drystone walling might have similarities with the move towards hidden burial noted in Britain (Cauwe 2004, 219–220). At a similar date in Ireland, the evidence is more varied. Some of the sites with Middle to Late Neolithic dates, such as Ballymintra Cave, Co Waterford, are interpreted by Dowd (2015, 105) as intermediary period sites used in secondary burial rites. By contrast, Kilgreany Cave, also in Waterford, seems to have been used for successive inhumation throughout the period (Dowd 2015, 101). Weiss-Krejci's (2012, 127–131) survey of the evidence from the Iberian Peninsula also shows a variety of rites in use in the period around 3300–2900 BC, and a similar range of rites were still being practiced in France (Beyneix 2012; Vander Linden 2006). It should also be noted that, unlike in Ireland and

Britain, where most human remains from caves were earlier in date, cave burial is actually most common in the centuries around 3300–2900 in Belgium and Spain, and it continued to be extremely common in France.

Late Neolithic

There are mid-third-millennium BC radiocarbon dates on human bone from three sites, which suggest that the practice of hiding the intermediary period away in the depths of active caves continued into the Late Neolithic. North End Pot, North Yorkshire (Appendix 1, number 33: NGR SD 6830 7653) is an open fissure in the limestone pavement on North End Scar near Ingleton. The fissure leads to a vertical shaft which descends into a steeply sloping rift and, ultimately, at

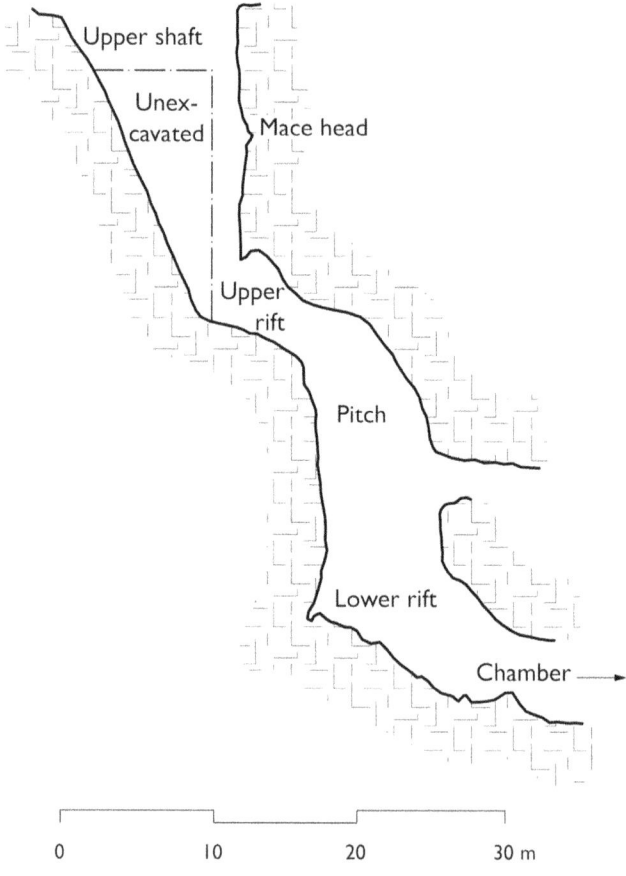

7.7 Sketch section of the entrance shaft and rift at North End Pot, showing the approximate position of the human remains (after Leach 2006, Figure 3.1.2.1).

a depth of around 45 metres from the surface, into a large chamber known as 'St George's Hall' (Leach 2006, 51 and see Figure 7.7). The exploration of North End Pot was carried out in the 1980s by Earby Potholing Club, and the human remains are now in the Lord collection at Lower Winskill. The site was clearly used for burial in the Iron Age. A human cranium recovered from the upper fills of the entrance shaft was radiocarbon-dated to rule out the possibility of a police enquiry. This established that there was some burial activity at the site in the Late Iron Age and that the entrance shaft was largely filled with sediment by this date. (Leach 2006, 52). The only artefact discovered during the excavations was a Late Neolithic antler mace head found on a narrow ledge in the shaft wall, approximately 11 metres below the surface (Gilks and Lord 1993, 57).

The human skeletal material from North End Pot was analysed by Leach (2006) as part of her study of the sites in the Lord Collection. She was able to divide the sixty-nine fragments recovered into two groups based on the reported context where they were found and by differences in their condition. Nine fragments from the upper shaft and entrance area are parts of two different sub-adults, a child of 4–5 years old and an adolescent of between 14 and 18 years old. These two burials certainly took place in the Late Iron Age, based on the radiocarbon dating carried out for the police. The rest of the assemblage came from deeper within the cave and represents part of at least two adults. There were twelve fragments from the lower fill of the entrance shaft; six were found on the floor of the upper rift and a further forty-two on the floor of the lower rift (Leach 2006, 54–55). A mandible fragment from the lower shaft area was submitted for radiocarbon dating as part of Leach's project (Appendix 1: OxA-14265: 4176 +/- 31 BP), and this result would give a date which calibrated to between 2885 and 2635 BC at two standard deviations.

Leach (2006, 61–63) identified four possible examples of perimortem fractures in the assemblage from the lower parts of the cave. Although the amount of bone surviving was small, there were two metatarsals in the assemblage. Taking into account varying patterns of weathering on the bone and the absence of evidence for animal modification, Leach (2006, 57–67) considered two possible mechanisms for the distribution of bone shown in Figure 7.6. The shaft may have been partially open during the Late Neolithic and acted as a natural trap, with the perimortem fractures caused by falling injuries. Alternatively, the fleshed bodies may have been successively inhumed on whatever stable surface existed in the upper shaft in the Late Neolithic. The bones would then have been transported by gravity and scree movement deeper into the rift. Despite the presence of the perimortem fractures, which could of course have been a cause of death, Leach (2006, 65) considered that the most plausible interpretation of the assemblage was that successive inhumation took place within the upper shaft. This may have been considerably deeper than it is today, possibly as deep as the ledge where the antler mace head was discovered.

Orchid Cave, Denbighshire (Appendix 1, number 36: NGR SJ 2002 6062) was explored by North Wales Caving Club in 1981. The site is described as a descending passageway around 13 metres long (Ebbs 2017), and a catalogue of the finds was published by Davies (1981). These included animal bones as well as human remains and a worked bone toggle which is likely to be Iron Age in date (Guilbert 1982). The human and animal bone, together with a single flint scraper, came from a small chamber at the end of the passage (Ebbs 2017). Davies' (1981) assessment of the human bone suggested that there were a minimum of three individuals present: two adults and a sub-adult. Among the elements he noted in his catalogue were a patella and seven hand and foot bones. He also noted the presence of rodent gnawing on the ends of some of the long bones. Further human and animal bone survives in Orchid Cave (Ebbs 2017). The radiocarbon date is on a pelvic fragment (Appendix 1: OxA-3817: 4170 +/- 100 BP) and, at two standard deviations, it would calibrate to between 3010 and 2470 BC. It is likely that the funerary rite at Orchid Cave was successive inhumation, but it is not clear whether this originally took place near the surface or if the bodies were brought to the underground chamber for burial.

Blue John Cavern, Derbyshire (Appendix 1, number 8: NGR SK 1319 8320) is a large and intricate underground system that was formerly part of a complex of dolines, although it is now accessible through historical mineworkings. New explorations of part of the system in 2010 discovered human and animal bone in a boulder choke which probably connects to a visible doline on the surface (Nixon 2011, 93–95). The only human bone in the assemblage was a midshaft fragment of adult right tibia. In view of the fact that this was a single find within a complex and highly active system, no attempt at reconstructing a funerary rite, or even an original place of deposition, was possible. A radiocarbon date on this bone (Appendix 1: GU-21803: 4125 +/- 40 BP) would calibrate to between 2870 and 2580 BC at two standard deviations.

In contrast to some areas of continental Europe, particularly Belgium, France and the Iberian Peninsula, where cave burial is extremely common throughout the third millennium BC, there are very few definitely Late Neolithic cave burial sites. This is consistent with the evidence from the British and Irish Late Neolithic more generally, where inhumation burial was rare (Cummings 2017, 192–193). In view of the increasing evidence for cremation burial from monuments, for example the Sarn-y-Bryn-Caled 2 ring ditch, Powys (Gibson 1994, 161), it may be that some undated cremation burials from caves in Britain are also Late Neolithic. In Ireland, Late Neolithic human remains from caves are also very rare, with only the possible activity at Ballynamintra, Waterford, discussed here falling into this period. Despite the low number of sites, it appears that a successive inhumation rite was the usual one. As in the Middle Neolithic, the intermediary period was probably something that was supposed to take place away from the world of the living. The material and embodied narrative represented by the decompositional changes to the body was hidden by placing the dead deep into caves and shafts. The actions of caves also took place away from the world of the living. As has been discussed here for the Middle Neolithic, sites such as Blue John Cavern and North End Pot show that cave sedimentation and flow processes would have had a significant impact on the temporality of the intermediary period. There is, however, a paradox here. The actions of caves and bodies would have contributed much more to the funeral process than in the earlier Neolithic sites discussed in Chapter 6, but, because they were largely concealed in deeper parts of the cave systems, they would have been much less integral to human understandings of the rite. Perhaps this was the point, that cave actions were supposed to be incomprehensible or obscure, providing a directed path or journey for the dead which was beyond human agency. There were some continental Late Neolithic burials which seem to show similar concerns, for example, the use of dry stone walling or cists to hide the body in French and Belgian Final Neolithic cave burial sites (Roussot-Larroque 1984, 160; Cauwe 2004, 219–220). However, the analogy must not be stretched too far; these European sites are a sub-set of a much greater number of burial caves with, especially in France, a wider diversity of burial practice.

Early Bronze Age

The sites discussed in the preceding section were all used during the Late Neolithic, but they do not have evidence for continued use across the transition into the Early Bronze Age. There are, however, at least four British caves with radiocarbon dates on human bone which, when calibrated, cover the period around 2400 BC. This date marks the probable transition between the Late Neolithic and the Early Bronze Age in Britain (Cummings 2017, 234). The human bone discovered in Mother Grundy's Parlour, Derbyshire (Appendix 1, number 31: NGR SK 5358 7426) has been radiocarbon-dated on two different occasions, giving a range of results from the Late Neolithic to the Iron Age. Only one of these dates is now regarded as reliable, and this would place the death of one of the individuals in the Early Bronze Age (Hedges et al. 1996, 396). The site itself is a large rock shelter on the north side of Creswell Craggs, and it has had a long excavation history which is summarised by Campbell (1977, 60–62). Most of the human bone from the site was discovered during excavations in the late 1870s by Boyd Dawkins and Mello and was described as coming from fragmentary juvenile remains inside the rock shelter. The adolescent molar dated by Hedges and colleagues (1996, 396: Appendix 1: OxA-4442: 3720 +/- 80 BP) was discovered in 1959 by Charles McBurney outside the cave in the spoil heap from earlier excavations (Campbell 1977, 62). Calibrating this result at two standard deviations would give a date of death for this individual between 2430 and 1895 BC.

A similar radiocarbon date comes from Ash Tree Cave, Derbyshire (Appendix 1, number 2: NGR SK 5148 7615), which is a medium-sized chamber, 2.7 metres wide by 4.8 metres deep, on the north-west side of Burntfield Grips. It was excavated between 1949 and 1957 by Leslie Armstrong (1956). He describes the upper deposit in the cave as a clast-supported scree deposit around 0.5 metres deep. This layer had many large limestone blocks within it, which had probably eroded from the roof. At the base of the scree, towards the entrance, there were indications that some of these blocks had been piled together to cover human remains. Armstrong (1956, 57–58) describes a deposit which included the remains of at least four different individuals who appeared to have been successively inhumed. Skeletal elements described as being present include phalanges and most of the major skeletal elements, although no crania were recorded. At the back of the entrance chamber, a fissure opened into a blocked passage. When this passage was excavated, a limestone cist was discovered around 6 metres from the cave entrance. This cist included the remains of at least two further individuals. Armstrong (1956, 59) was of the opinion that after the cist was constructed, the passage had been deliberately blocked before the deposition of the four bodies in the entrance chamber. A juvenile distal left tibia fragment from the deposit by the cave entrance was radiocarbon-dated (Appendix 1: OxA-4446: 3730

+/- 90 BP). This result would calibrate, at two standard deviations, to between 2460 and 1915 BC.

The very similar results from both Ash Tree Cave and Mother Grundy's Parlour might suggest that successive inhumations in cave mouths and rock shelters continued to take place at least until the very end of the Neolithic. Other sites which may span the transition from the Late Neolithic to the Early Bronze Age show that burial in deeper systems, or burials which drew upon the active nature of large systems to move bodies deeper underground, were also taking place at this date.

One of these sites is Ashberry Windypit 1, North Yorkshire (Appendix 1, number 4: NGR SE 5709 8501). The site is part of a group of fissures, known as the Ryedale Windypits, on the North York Moors, near Helmsley. The windypits are mass-movement fissures created by the slippage of rocks rather than the erosion of water. As such, they are complex and maze-like both in plan and section (Leach 2006, 251 and see Figure 7.8). Most of the archaeological material from the site was removed during excavations in the 1950s and 1960s, and the archive and finds have been re-assessed in detail by Leach (2006, 225–229). In addition to the Early Bronze Age finds, Romano-British material was also discovered during the exploration of chambers B and C (Leach 2006, 227). The human remains came from two levels of the deepest chamber, chamber D, where they were

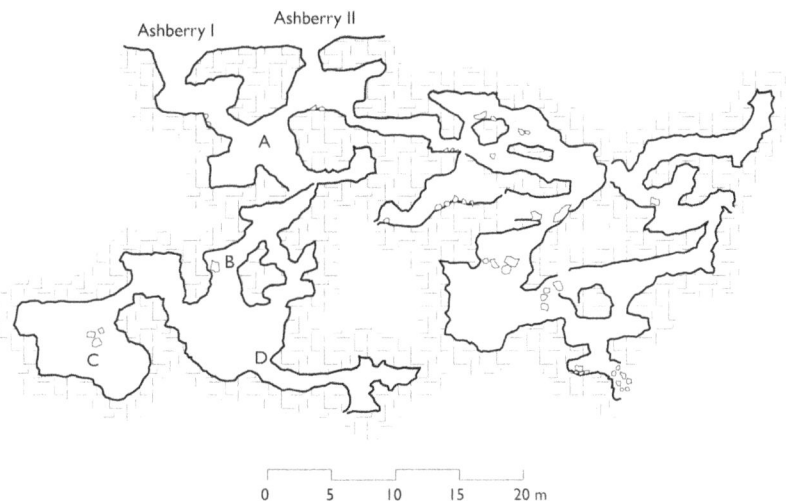

7.8 Simplified section through Ashberry Windypits I and II, showing the location of the Late Neolithic/Early Bronze Age human remains in chamber D (after Cooper et al. 1976, Figure 12; and Leach 2006, 251).

associated with animal bone, Beaker pottery, worked stone and a bone pin. Leach (2006, 226) considered that this assemblage was the remains of one or more burials which had originally been deposited in the upper part of chamber D and had become disarticulated as the sediments moved downslope. The skeletal material from Ashberry Windypit I has had a complex curation history, but Leach (2006, 230–232) was able to confidently identify a minimum of three individuals in the assemblage from chamber D: one adult male, one adult female and a sub-adult (Leach 2006, 235).

Radiocarbon dates were obtained on the two adults (Appendix 1). Although these dates are similar, they are statistically unlikely to represent a single event, and therefore the dates have been modelled in OxCal 4.3 (Bronk Ramsey 2009; Reimer et al. 2013) on the assumption that there were repeated burial events. This would suggest, at two standard deviations, that the first burial took place between *2460* and *2215 BC* and the last burial between *2300* and *2060 BC*.

Leach (2006, 240) interpreted these burials as successive inhumations which were accompanied by the stone tools and Beaker pottery. They were probably placed at the top of chamber D, rather than moving from the surface. As may have been the case at the earlier site of Totty Pot, this would have involved some relatively difficult manoeuvring to get the bodies and the accompanying artefacts to this location (see Figure 7.8). She notes that, unusually for a cave assemblage, the radiocarbon results and the expected date of the artefacts accompanying the burials would broadly agree. In particular, the Beaker pottery from Ashberry Windypit I is an example of an All Over Corded Beaker, which would probably date to around or just before Needham's (2005, 206) 'fission horizon' between 2250 and 2150 BC.

There are two very similar late third-millennium BC dates on human remains from Charterhouse Warren Farm Swallet, Somerset (Appendix 1, number 16: NGR ST 4936 5457). The Late Neolithic and Early Bronze Age archaeology of the site was discovered during excavations between 1972 and 1976 (Levitan et al. 1988). The site is a doline which led to a vertical shaft at least 22 metres deep. A side passage from the upper part of the doline leads to a cave system. This system was explored separately between 1983 and 1986. Radiocarbon-dated human remains and artefacts show that this side passage and cave system were accessible until the Roman period (Levitan and Smart 1989, 393–394). Beneath the upper 6 metres of deposit, which were removed without archaeological recording during initial cave exploration, the doline shaft fill could be divided into four

Deep time

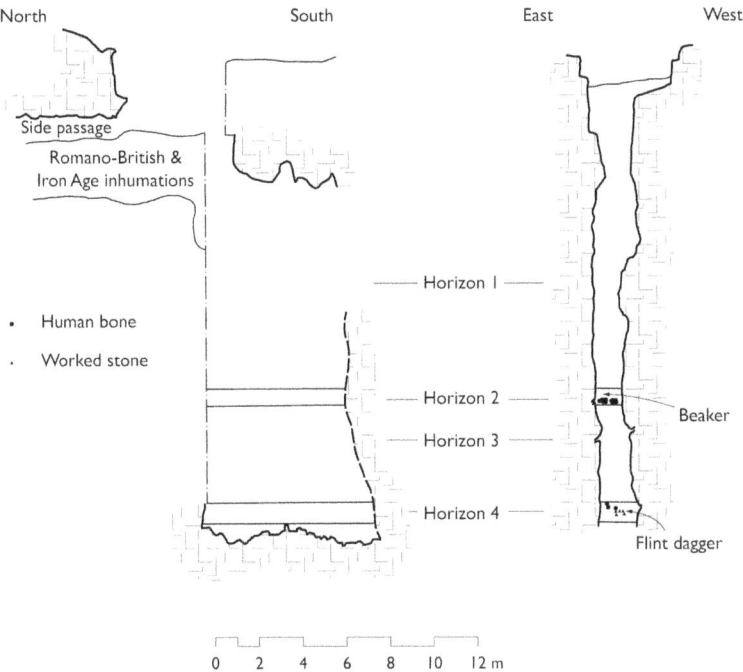

7.9 Sections through the excavated parts of Charterhouse Warren Farm Swallet (after Levitan and colleagues 1988, Figures 5, 6 and 8).

successive layers (Levitan et al. 1988, 200–202 and see Figure 7.9). The uppermost of these, horizon 1, was a very coarse clast-supported scree with many large angular limestone boulders. This layer was approximately 8.9 metres deep. It was clearly forming during later prehistory, as a butchered aurochs horn core from horizon 1 gave a radiocarbon result (BM-731: 3247 +/- 37 BP: Levitan and Smart 1989, 391) which would calibrate to between 1625 and 1440 BC at two standard deviations. At the base of this layer was horizon 2, a layer of silty clay loam approximately 0.7 metres thick, which contained large quantities of bone, both animal and human. Also within horizon 2 was a cluster of sherds from an S-profiled Beaker. Below this layer was horizon 3, a layer comprised of smaller angular limestone fragments and silty clay lenses, which was 4.45 metres thick. From horizon 3 came three abraded sherds of Grooved Ware and some animal bones, but no other archaeological material. At the limit of the excavated deposits was horizon 4, a 0.7-metre-thick deposit of silty clay. Finds from this layer included neonatal and infant human bones, worked antler, a bone pin, animal bone and worked flint, including an extremely fine dagger (Levitan et al. 1988, 201). The fill of the shaft was interpreted as a largely natural series of events. Freeze/thaw erosion of the limestone was probably responsible for the formation of horizons 1 and 3, with horizons 2 and 4 formed from periods of more active soil erosion around the mouth of the shaft (Levitan et al. 1988, 199).

The artefactual evidence from the lower layers of the shaft fill consists of a mixture of diagnostically Late Neolithic finds, such as the Grooved Ware and the bone pin, and later objects such as the dagger and the Beaker pottery (Levitan et al. 1988, 206–207). The two radiocarbon dates (Appendix 1) come from human bone from horizon 2 (OxA-1559) and horizon 4 (OxA-1560). These two dates can be modelled in OxCal4.3 (Bronk Ramsey 2009; Reimer et al. 2013) on the assumption that the bones concerned were originally deposited in the layers where they were found, and that they therefore represent a series of different burial events. On this basis, burial in horizon 4 started between *2455* and *2150 BC*, with burial in horizon 2 dating to between *2400* and *2130 BC*. Given the extremely similar nature of the dates and the range of artefact associations, the alternative possibility was considered that the bone represents a single burial event which has been redeposited in two different layers of the shaft fill. Statistically, the two dates could represent a single burial event. If they were combined, then the two results would suggest that, at two standard deviations, burial took place between *2265* and *2035 BC*.

Most of the human bone from Charterhouse Warren Farm Swallett comes from horizon 2, with a small number of fragments from horizon 4. The published estimates for the minimum number of individuals from the site are high; Chamberlain (2014) suggests there were more than thirty. However, this figure includes the later prehistoric and Romano-British burials near the surface and other later prehistoric human bone from the passage cave. Levitan and colleagues (1988, 213–214) provide an outline catalogue of the bone in each layer without providing a formal estimate for the minimum number of individuals from the entrance shaft. However, on the basis of their descriptions, there was both a neonate and a slightly older infant from horizon 4. There were also infant and neonatal bones in horizon 2, but there were also at least two older juveniles and two adults. Therefore, if all the bones from the lower entrance shaft are treated as a single group of late third-millennium BC burials, the absolute minimum number of individuals would be six. This is almost certainly an underestimate, but, given that the total number of skeletal fragments recorded for both horizon 2 and horizon 4 is only 228 (Levitan et al.

1988, 210), probably not by much. Some of the bones are described as having cut marks around the points where they would articulate (Levitan et al. 1988, 201). The bones from horizon 2 listed in the catalogue as having cut marks were six cranial fragments and one scapula, although a cut-marked humerus is also illustrated. Some of the cranial fragments and teeth were also described as burnt (Levitan et al. 1998, 212). The only other possible example of people actively defleshing the corpse like this during the intermediary period comes from the Late Neolithic site at North End Pot described here. At that site, Leach (2006, 63) tentatively identified a perimortem cut mark on a humerus, although, because of extensive abrasion of the bone, this was not a definitive identification. At Charterhouse Warren, all parts of the skeleton appear to be well represented for both the adults and the juveniles from horizon 2. There are phalanges, metacarpals and metatarsals, all the major long bones, pelvic and spinal elements and both crania and mandibles. The infant bones in horizon 4, although much fewer in number, also include one phalanx, long bones, axial elements and a cranial fragment.

Funerary practices at Charterhouse Warren Farm Swallet were clearly complex, and it is possible to interpret the cave sediments and the human bone evidence in a number of different ways. In their original report, Levitan and colleagues (1988, 232) interpreted the bodies and artefacts as a series of placed deposits within the shaft. They assumed that the bones had been disarticulated prior to deposition. They considered the possibility that artefacts and bones had eroded from the shaft entrance into their final positions but concluded that both bone and artefacts were too well preserved for this to have been the case. They interpreted the shaft sediments as showing the infill to have been an entirely natural process. In the light of the radiocarbon dating evidence, which suggested that the material in horizons 2 and 4 were broadly the same date, they subsequently modified this interpretation slightly. They proposed (Levitan and Smart 1989, 393) that horizon 3 was actually a deliberate dump of material, added between two burial episodes of the same late third-millennium BC phase. However they entered the shaft, the human remains had been part of an intermediary period before they did so. The cut marking and burning of some of the bone does suggest active human involvement in the process of disarticulation. On the other hand, it is likely, from the range of skeletal elements present, that the bodies were still at least partly fleshed when they were deposited.

I think that there are three possible explanations for the range of evidence from this site. Artefacts and dismembered parts of bodies may have been brought to the site and deposited in horizons 2 and 4 as suggested here, with horizon 3 as a deliberate dump. Alternatively, the shaft infilling may have been entirely natural, as originally suggested, and the deposits of human bone and artefacts may have represented curated and associated assemblages which were placed in the shaft on separate occasions. It may be that people were excavating them from more 'conventional' Beaker funerary sites. Lastly, it is possible that the unexcavated portion of the shaft conceals another side passage, and that horizons 2 and 4 are made up of material which has eroded from this area. In this case, there may have been a single Beaker burial episode in this hypothetical part of the system, which was eroded on successive occasions to produce two widely separated deposits with very similar dates. Whichever of these explanations is the correct one, the burial rite at Charterhouse Warren Farm Swallet would have involved a relatively short intermediary period ending in the dismemberment of the still partially fleshed body, followed by final burial a long distance underground.

As in the Late Neolithic, the last quarter of the third millennium BC saw relatively few burials in caves. There seems to have been a large degree of continuity with earlier cave burial practices, although Beaker period material culture was clearly being drawn upon. However, there was clearly a significant difference in burial practice at other kinds of sites by this date. So, although the few cave burials we currently know about show clear links with Late Neolithic cave burial rites, the connections to wider Beaker burial practice is not obvious. In particular, despite the overall increase in recorded burials in the period, there was no corresponding increase in cave burials. Beaker cave burials also do not seem to have been placed into dug graves or pits in the way that more conventional Beaker burials were. However, as Cummings (2017, 250) and Fitzpatrick (2011, 200) have pointed out, 'standard' Beaker graves also have elements of continuity with earlier burial practice. The most important of these, which bears directly on the evidence discussed here for the manual disarticulation of bodies, are those burials which show that an intermediary period was an important part of Beaker funerary rites. At the recently excavated collective grave at Boscombe in South Wiltshire, at Chilbolton in Hampshire and at Manston in Kent, Beaker burials show evidence that the bodies were being re-arranged and the graves were being re-opened after an extended intermediary period (Cummings 2017, 250).

Although cave burials in the Beaker period continued to use deep and inaccessible locations in a similar way to those of the Middle and Late Neolithic, they are unusual in that they appear to be consistently associated with grave goods. Almost all of the other burials I have discussed in this book were not associated with objects of the same date. Part of this was undoubtedly because the movement of human remains by cave processes has destroyed any relationships that did exist between burials and artefacts. However, the consistent survival of what seem to be grave goods in the Beaker period must reflect a change in burial practice. This use of grave goods is the most obvious connection between Beaker cave burial practice and Beaker burial in flat graves and barrows. It is also important evidence for how the agency of objects, caves and bodies worked together in the construction of the intermediary period at this date. There were two primary contributions from the cave environment. It would have provided an index of the passage of time through the movements of cave sediments and bodies, for example in the way that disarticulating human remains were apparently reworked and redeposited at Charterhouse Warren Farm Swallett. The relatively fixed overall form of the cave would also have provided long-term connections to ground the circulating references to earlier and contemporary cave burials. The significant quantities of material culture in these burials must indicate a close relational link – for the first time since the Early Neolithic – with other Beaker burial rites at other kinds of site. Despite the continued use of deep locations, it also seems as if there was a considerable amount of input from living people into these rites. Bodies seem to have been defleshed or disarticulated, and the survival of the pottery at both Charterhouse Warren and Ashberry Windypit 1 might suggest that the assemblages of grave goods were being actively curated during the intermediary period.

In western parts of mainland Europe, the relationship between caves and other kinds of Beaker burial site was slightly different. As discussed in Chapter 2, overall numbers of cave burials in Spain, France and Belgium did increase in the Late Neolithic and Early Bronze Age. Interestingly, Vander Linden's (2006, 324) discussion of Beaker burial in southern France shows that cave burial was much more integral to general mortuary practice there. As with the British cave evidence, southern French Beaker burials are often collective, have evidence for rites involving an intermediary period and used caves, hypogea and dolmens as burial spaces. In another parallel with the British caves, general Beaker burial practice in this region is largely distinguished from Late Neolithic burial by the increasing use of grave goods and by an increasingly standardised repertoire of objects in graves. The evidence for disarticulation and defleshing of bodies at Charterhouse Warren Farm Swallet is echoed in the complex treatment of the dead seen at some Belgian Late Neolithic sites (Cauwe 2004, 220) and in the Iberian Peninsula (Weiss-Krejci 2012, 129–130). Cut-marked bone has been identified at several Spanish and Portuguese sites, and it is suggested that bones were manually disarticulated before being partially burnt. In some cases, as noted in Chapter 6, this treatment may have been applied to groups of bones which had been curated for considerable periods of time.

Conclusions

During the Middle and Late Neolithic, from the last part of the fourth millennium until the last quarter of the third millennium BC, funerary rites in British caves seem to have formalised. Although there were less caves in use during this part of the Neolithic, there is a degree of consistency about the evidence which contrasts with the diverse range of practices described in Chapter 6. Wherever there is good evidence for burial practice in the Middle and Late Neolithic, that funerary rite seems to have been successive inhumation. This would have meant that human agency was no longer a central part of the intermediary period. The material narrative of change, which would presumably have allowed people to gauge the appropriate time for the end of the intermediary period, was now something which depended on the actions of animals, caves and bodily decomposition. There is also good evidence that not only were living people not actively involved in the intermediary period, but also that it was something that should be physically separated from their lives. Cave burials took place deeper into systems or were otherwise separated from other activities. It is also possible that less people were being buried in each cave by the Late Neolithic. Comparing all fifteen caves where we have good evidence for successive inhumation in any period, we can see that the mean minimum number of individuals (MNI) surviving from each site in the Early and Middle Neolithic is similar. The figure for the Early Neolithic is 4.13, and for the Middle Neolithic it is 5.00. The comparable figure for the Late Neolithic is a mean MNI of only 1.75. However, both the Early and Middle Neolithic figures include one site each (Hay Wood Cave and Backwell Cave respectively), with significantly higher numbers of burials. The Late Neolithic data comes from only four sites, and so this pattern should probably be regarded as suggestive rather than conclusive at this stage.

The very few Beaker-period sites with good evidence show that both successive inhumation and the desire to spatially separate the burials continued after the end of the Neolithic. However, the evidence from Charterhouse Farm Warren Swallet shows that the tradition of the living not engaging with the decomposing corpse did not. Beaker cave burials seem to have come to involve the practice of manual disarticulation and re-arrangement of the corpse. As we have seen, this is something that can be paralleled both in Beaker flat graves and in European Beaker burial practice. Therefore, by the Beaker period, it seems as if what had become a relatively uniform cave burial rite was once more open to influences from other contemporary burial practice. Reviewing the evidence from the beginning to the end of the Neolithic over the last three chapters of this book, I have given primacy to the specific evidence from individual caves. In the final chapter, I will attempt a broader overview of how the temporality and the human and natural agency of death and burial influenced cave burial practice throughout the period.

8

Temporality, structure and environment

Introduction

In this final chapter, I will attempt to provide a synthetic overview of the evidence considered in detail in the previous three chapters. Throughout this book, I have worked on the assumption that we can best understand multi-stage collective burials by understanding the workings of the intermediary period. I have adopted Hertz's (1960, 201–202) insight that the intermediary period connects the physical condition of the decomposing corpse with the changing social role of the deceased. The *soul*, for want of a better word, exists in a liminal state during the intermediary period as the surviving kin negotiate all the complexities of grief, unfulfilled obligations and unpayable debts which have been occasioned by the death. The temporal congruence between the social and biological transformations concerned allows the state of the corpse to act as an *index*, in Gell's (1998, 23–27) sense of the term, for the state of the soul. Additionally, as Harris and Robb (2012, 674) have pointed out, human bodies undergoing this transition are one example of the way in which the ontology of the body can be multimodal. A human body, especially a corpse going through the points of transition within a network, can act and be conceptualised as both a person and a thing. Therefore, the changes happening to dead bodies would have also served to highlight the multimodal nature of the ontology of the body.

It is important to remember that the intermediary period does not only involve the dead body and the mourners. The experience of the intermediary period also derives from the interaction between the agency of living people, the agency of the decaying corpse and the agency of the environment in which the intermediary period takes place. Following the discussion of the agency of inanimate things in Chapter 4, I have taken the position that all three of these kinds of agent would have an equal role in constituting the kind of intermediary period which takes place. Therefore, the most effective way to understand the different kinds of intermediary period and the different kinds of Neolithic collective burial practice in caves is through the study of their related material traces. The bodies, caves, sediments and artefacts in and around the burials would have provided the material clues which allowed Neolithic people to reconstruct and understand the intermediary period and, through that, to comprehend the processes of death and the progress of the soul after death. These physical indices of change are the events which constitute the temporality of the intermediary period.

Temporality

The first aspect of cave burial that I wish to explore at a national scale picks up on these discussions of time and temporality from Chapter 4. The modelled chronologies for those sites which have multiple radiocarbon dates show that there is great variability in the duration of burial activity. Some sites, such as Hay Wood Cave, appear to have been used intensively for a relatively short time. Others, such as Raschoille, which has a similar number of overall radiocarbon determinations, seems to have been used much more episodically over a longer period. These differences in the intensity and duration of use are likely to have reflected the temporal cues provided by the social networks around mourning, the caves and landscapes in which burial took place and bodily decomposition. All of these things in turn would have been a vital part of the way in which specific rites were remembered and reproduced. Therefore, investigating the temporality of particular cave burial rites integrates many different scales, from the likely time taken for each part of the multi-stage rite to the overall duration of that kind of funeral practice at a national scale. In the preceding chapters, I have already established an outline chronological model on the basis of the dates from individual sites. This suggests that Early Neolithic practices were the most diverse, with successive inhumation, secondary burial, primary burial, a cult of the head and possible mummification all taking place. In the Middle and Later Neolithic, successive inhumation burials seem to have been the norm, with no solid evidence for any other practice. Where there are multiple radiocarbon dates from different sites, it is possible to refine this outline model.

In the case of the burials of isolated skulls from the Pennines discussed in Chapters 5 and 6, the modelled duration of the practice is shown in Figure 8.1. With the exception of the single possible example of mummification, this is the least securely dated of the burial rites, with only four dates from three different sites. If the two dates on

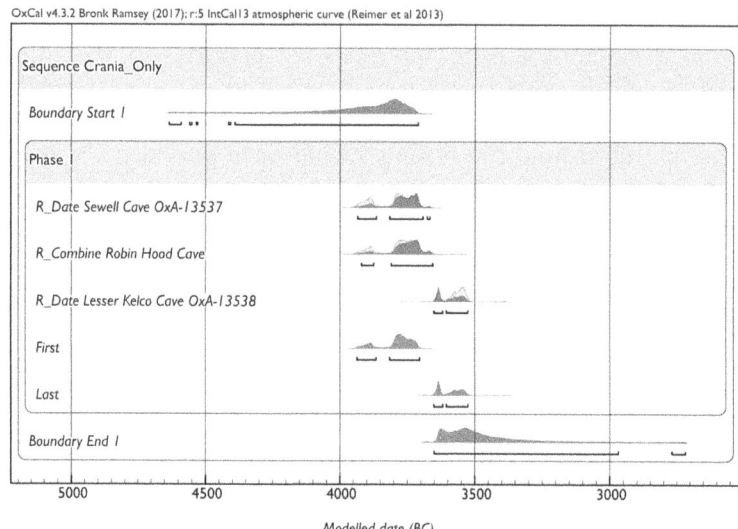

8.1 Duration of the 'head cult' burial rite in the Yorkshire and Derbyshire Pennines. See Chapters 5 and 6 for the original published sources for these dates.

what I have assumed is a single skull from Robin Hood's Cave are combined, the result is very similar to the single date from Sewell's Cave. Treating all three sites as examples of a related practice of secondary burial of the head, I would suggest that this rite was short-lived and belonged at the beginning of the local Neolithic. At two standard deviations, the earliest burial took place between *3940 and 3705 BC* and the latest known burial dated to *3655–3525 BC*. The modelled date for the start of this practice is therefore very similar to Griffiths' (2014a, 20) estimate for the beginning of the Neolithic in Yorkshire and Humberside. Given the low number of burials involved, that there were only ten individuals from all three caves and that only three of those have been dated, it is entirely possible that the practice persisted longer than suggested by these dates. However, as the non-cave examples of skulls listed by Schulting and Wysocki (2005, 128–129) are all likely to belong to the Earlier Neolithic and no further examples of cranial secondary burial have been identified in the forty-eight sites in this study, I feel confident that this practice did not persist into the Middle Neolithic.

The temporality of an individual burial in this tradition can be best understood by referring to the taphonomic research reviewed in Chapter 3. None of the skulls from any of the sites show cut marks to the mandible or cervical vertebrae, but in two cases – at Robin Hood's Cave and at Sewell's Cave – the temporo-mandibular joint seems to have remained articulated. This can happen relatively late in the disarticulation process (Knüsel 2014, 34), so rather than seeing these heads being cut off a recently deceased body, we should imagine desiccated but still articulated corpses being manipulated to remove the head. On the other hand, the presence of the third cervical vertebra at Robin Hood's Cave might indicate that the intermediary period in that case was short or, more probably, that the head was removed at or just after death (Randolph-Quinney, personal communication). The lack of reported evidence for canid gnawing shows that the bodies were probably buried during the intermediary period. The Wataita example discussed in Chapter 3 can be used as an analogy. In this case, the bodies were sufficiently decomposed after an intermediary period of around 24 months that only the crania were removed (Kusimba et al. 2005, 250). The total intermediary period in most of the British Neolithic examples was probably only between 6 and 12 months before the heads were removed and given secondary burial.

Secondary burials which also involved post-cranial bones are also quite poorly represented in the overall sample of sites. Figure 8.2 shows the probable overall duration of this funerary rite, using only those three sites where there is reasonably strong evidence for secondary burial: George Rock Shelter, Broken Cavern, and Chelm's Combe. As these sites form a coherent regional group, I would argue that, as with the

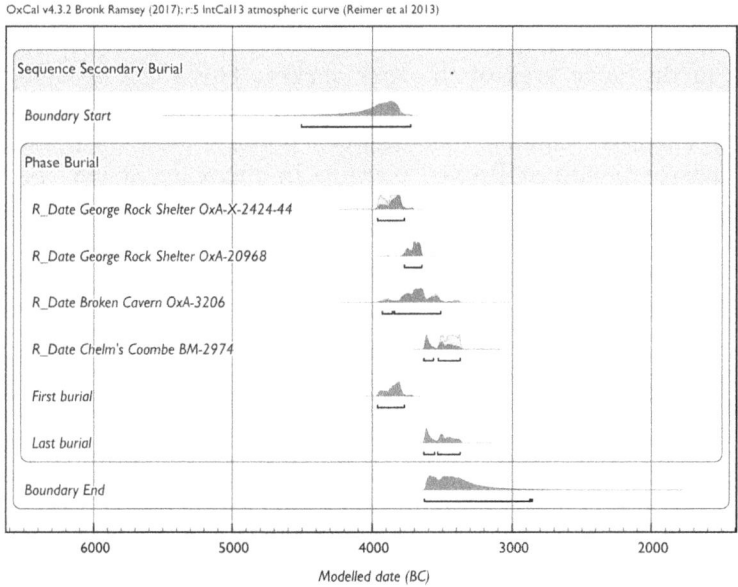

8.2 Probable duration of secondary burial rites in South Wales and south-western England. See Chapters 5 and 6 for the original published sources for these dates.

head burials considered here, this practice was relatively short-lived and belonged at the beginning of the local Neolithic. At two standard deviations, the earliest burial took place between *3965* and *3770 BC* and the latest known burial dated to *3635–3370 BC*. The modelled date for the start of this rite can be compared to the estimates for the beginning of the Neolithic in south-west England, *3940–3735 BC*, and South Wales, *3765–3655 BC*, provided by Whittle and colleagues (2011, 516–517, 548). This would suggest that secondary burial in this region is another exclusively Early Neolithic rite. This suggestion must remain slightly tentative until more dated individuals have been identified. There are ten undated individuals from Chelm's Coombe and six from George Rock Shelter which could be used to help resolve this issue.

Secondary burial would have been different from other funerary practices because it would have had a distinctively different temporality. It would also have offered the opportunity for living people to interact more extensively with the decaying corpse during the intermediary period. Perhaps for this reason, we can see evidence that pottery, lithics and animal bone were all being placed with the bodies during the extended secondary burial process. We can follow this engagement between objects, bodies, living people and the cave. We have evidence for the location and duration of the intermediary period at all three sites. At George Rock Shelter, we know that the intermediary period lasted less than 3 years and seems to have taken place close to the rock-shelter wall. Fragments of several different Early Neolithic pots, along with large quantities of animal bone, were found in the same area of the rock shelter. This suggests that food and cooking vessels were being deliberately fragmented alongside the bodies. There is a similar link between pottery, food waste and the intermediary period at Broken Cavern. In this case, it may be that the 'midden material' (Roberts 1996, 203) was the remains of feasts which took place elsewhere as part of the temporality of the intermediary period. These feasts would have defined and marked the social transformations which were an integral part of this stage of the burial. Their direct connection to the rite is shown by the deposition of this material alongside the changing bodies within the cave. At Chelm's Combe, I have assumed that the intermediary period took place within the main rock-shelter area. Here again, there was evidence of Early Neolithic pottery being placed alongside the bodies as they decomposed. However, unlike at George Rock Shelter, some of these vessels were largely complete (Balch and Palmer 1926, 108–110).

The physical environment in which the intermediary period took place was also markedly similar at both George Rock Shelter and Chelm's Combe. In both cases, the bodies were easily accessible within open rock shelters. They were placed within and upon granular tufa deposits, which were probably actively forming while the burials took place. This would have had a significant effect on anyone who was interacting with the decomposing bodies and upon the bodies themselves. Anyone digging into the tufa, or even moving across the surface of it, would have rapidly become coated in a fine white dust in dry weather and plastered in thick grey marl if it was wet. This combination of open rock-shelter environment and tufa-rich sediment would have created an environment which allowed access to the bodies as they underwent the changes necessary for the intermediary period, but would also have sanitised the appearance of the corpse. Colour changes and odours would have been masked by the highly persistent bleaching effect of the granular tufa. Chelm's Combe is also the only site where there appears to be evidence for the final stage of a secondary burial rite. If the rock-cut chamber at that site is typical, then the pottery and feasting evidence which formed part of the intermediary period would have had no place in the final secondary burial. It was only the entirely clean bones that were placed in this final repository. It is possible that the final secondary burial for the human remains from George Rock Shelter took place at one of two nearby chambered tombs: St Lythans or Tinkinswood. Some of the human bone at least one chambered tomb in South Wales, Parc le Breos Cwm, seems to have been placed there in a secondary burial rite (Whittle and Wysocki 1998, 157–158). However, as discussed in Chapter 6, the nearest cave to this site with Neolithic human remains is Catole Cave. Despite being within a few hundred metres of each other, analysis by Schulting (2007, 592–593) seems to show that the bone from the cave and the chambered tomb came from separate burial populations.

The relative rarity of secondary burial rites from caves seems to mirror our current understanding of the rites used in Early Neolithic chambered tombs. As discussed in Chapter 3, recent taphonomic studies of chambered tomb human bone assemblages have come to the conclusion that, in the large majority of cases, the burial rite was a form of successive inhumation. However, there are also some examples of secondary burials from chambered tombs in the south-west. At Parc le Breos Cwm, this may have been the case for the bones from the chambers, although the passage deposits almost certainly represented successive inhumation (Whittle and Wysocki 1998, 158). At Pipton in the Black Mountains of Wales, Wysocki and Whittle (2000, 599–600) identified evidence for extensive re-arrangement and structured deposition of human bone after it had become skeletonised. This may represent something like the secondary burial at Chelm's Combe, where the whole extended process took place within the same relatively constrained space. The recent re-evaluations of Fussell's Lodge and Wayland's Smithy, Wiltshire, concluded that it was probable that some of the human bone at both these sites arrived in a disarticulated condition (Whittle et al. 2007, 107; Wysocki et al. 2007, 69), although the majority rite in both monuments was also successive inhumation. Using the data from these recent dating programmes, it is possible to compare the date of known secondary burials in chambered cairns with the cave examples discussed here. Figure 8.3 compares the cave

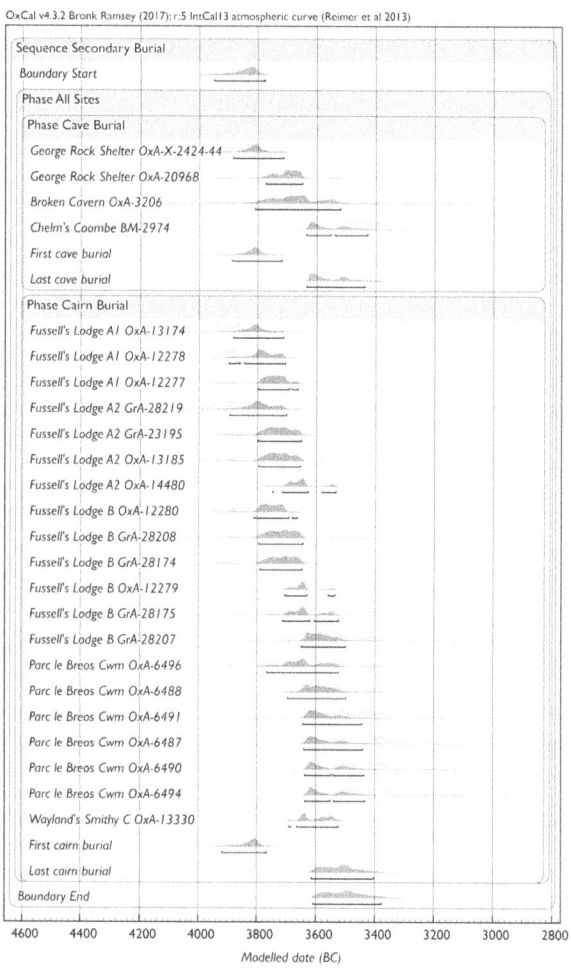

8.3 Dated examples of secondary burial from caves and chambered cairns in South Wales and south-west England. See Chapters 5 and 6 for the full publication details of the cave dates. The chambered cairn dates were published in Whittle and Wysocki (1998); Wysocki and colleagues (2007); and Whittle and colleagues (2007).

data with the dates of death of twenty individuals from chambered cairns. These are all people who seem to have received a secondary burial. This additional data would suggest that secondary burial was taking place in both caves and chambered cairns at the same date. The larger number of dates also allows us to refine the likely start for all secondary burial in South Wales and south-western England, which probably began between *3950* and *3780 BC* at two standard deviations.

As discussed here, the majority of caves in Britain have evidence for successive inhumation. Figure 8.4 shows the likely duration of this practice, based on the seventeen sites identified in Chapters 5, 6 and 7 as having good evidence for this rite. Unlike the more localised rites discussed here, some version of successive inhumation seems to have been used in all the cave regions of Britain. The likely start of this funerary practice lies between *3970* and *3780 BC* at two standard deviations. It is therefore very similar to the estimates obtained for the other, less well represented, burial practices beginning at some point in the first two centuries of the fourth millennium BC. However, unlike the various kinds of secondary burial, successive inhumations in caves seem to have continued throughout the whole of the Neolithic period. The modelled estimate for the end of this practice lies between *2400* and *2120 BC* at two standard deviations, with at least once good example, at Ashberry Windypit 1, being associated with Beaker period material culture. Although the majority of the individual radiocarbon results are early, most of the sites which have multiple radiocarbon dates seem to have been in use for a relatively long time. In particular, the likely span of use at Raschoille was between 195 and 530 years; at An Corran, it was even longer, between 915 and 1165 years; and at Totty Pot, burial took place for between 530 and 830 years. The only exception to this is Hay Wood Cave, where the dating programme carried out by Schulting and colleagues (2013, 16) showed that the cave was only in use for between 150 and 400 years in the early part of the fourth millennium BC.

The evidence for the long-term use of many cave sites for successive inhumation provides an interesting contrast with the temporality of successive inhumation in Early Neolithic chambered tombs. Recent re-analyses of chambered tomb assemblages and new dating programmes incorporating Bayesian modelling have critiqued the earlier idea that chambered tomb burial took place over a long period (Bayliss et al. 2007, 97–99, for example). At a number of these sites, burial has now been shown to be a relatively short-lived practice. The Wayland's Smithy sequence, as modelled by Whittle et al. (2007), provides a good example of successive inhumation burial in which

OxCal v4.3.2 Bronk Ramsey (2017); r:5 IntCal13 atmospheric curve (Reimer et al 2013)

Sequence Successive
  Boundary Start Successive Inhumation
  Phase Successive Inhumation
    R_Date Thaw Head Cave OxA-14264
    Phase Raschoille
      R_Date Raschoille OxA-8432
      R_Date Raschoille OxA-8431
      R_Date Raschoille OxA-8433
      R_Date Raschoille OxA-8441
      R_Date Raschoille OxA-8442
      R_Date Raschoille OxA-8404
      R_Date Raschoille OxA-8443
      R_Date Raschoille OxA-8434
      R_Date Raschoille OxA-8444
      R_Date Raschoille OxA-8435
      R_Date Raschoille OxA-8400
      R_Date Raschoille OxA-8399
      R_Date Raschoille OxA-8401
      R_Date Raschoille OxA-8537
    Phase Darfur Crag Cave
      R_Date Darfur Crag Cave OxA-V-2137-51
      R_Date Darfur Crag Cave OxA-V-2137-50
    Phase Cave Ha 3
      R_Date Cave Ha 3 OxA-13539
      R_Date Cave Ha 3 OxA-14266
    Phase Chapel Cave
      R_Date Chapel Cave OxA-V-2138-07
      R_Date Chapel Cave OxA-V-2138-09
    Phase Bower Farm
      R_Date Bower Farm OxA-V-2137-49
      R_Date Bower Farm OxA-V-2137-48
    Phase An Corran
      R_Date An Corran OxA-13594
      R_Date An Corran OxA-13552
      R_Date An Corran AA-27744
      R_Date An Corran OxA-13550
      R_Date An Corran AA-27743

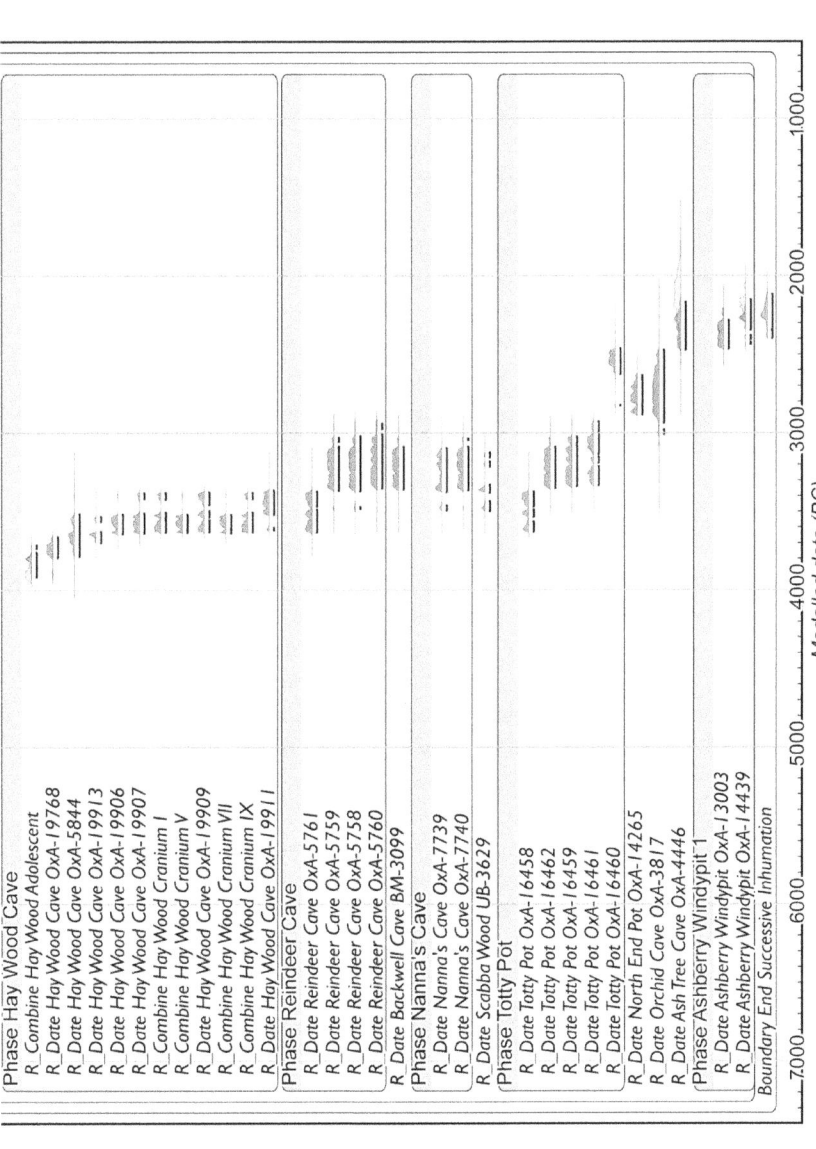

8.4 Modelled radiocarbon results from caves where there is good evidence for a successive inhumation rite. See Chapters 5, 6 and 7 for the original publication details of these dates.

the temporality of the rite can be relatively closely defined. At this site, all but one of the human bones sampled as part of the recent dating programme come from individuals who definitely had a successive inhumation burial. The only exception is the single date from individual 9, which was considered in the section on secondary burial here. Whittle and colleagues' (2007, 114–117) preferred model for the mortuary rites at this site begins with the deposition of successive inhumations in the timber mortuary structure of Wayland's Smithy I between *3610* and *3550 BC*. This structure was only in use for a maximum of 65 years, during which time the bodies of twelve people were successively inhumed in the tomb. There was then a period when no activity took place on the site, until the burial of the remains of the mortuary structure under the phase 1 mound between *3530* and *3435 BC*. There was then a further hiatus in the sequence before the construction of the much larger Wayland's Smithy II monument between *3490* and *3390 BC*. There was a further, less well preserved, episode of successive inhumation in the chambers of this monument, which lasted for between 1 and 185 years. Therefore, in contrast to the apparent pattern in most caves, successive inhumation at Wayland's Smithy was something that happened in relatively short phases, separated by times when the monument was not being used.

In part, this difference may be to do with the evidence which is available for the chronological modelling. In particular, the stratigraphy of a built structure such as a chambered cairn allows a clearer distinction to be made between events on the site. The two separate episodes of successive inhumation identified at Wayland's Smithy may have presented the appearance of a single prolonged deposit if they had been placed in a cave site. However, even allowing for some undetected breaks in the cave burial evidence, there still seems to be a significant difference in the temporality of cave successive inhumation. There are many more sites with evidence for prolonged use, and the practice of successive inhumation persists much later in the Neolithic in caves than it does in chambered tombs.

I think that we can see this difference as a reflection of the differences between the kinds of places which were being used for burial. Whittle, Barclay and colleagues (2007) have suggested that the generally rapid and episodic nature of successive inhumation in chambered long mounds was linked to the rapid and episodic nature of the construction of these monuments. Drawing on McFadyen's (2006) characterisation of this building process as 'quick architecture', the kind of successive inhumation which took place in chambered long mounds can be seen as similarly 'quick'. It may be that these intensive burial practices, and the monuments that contained them, were tied to managing the memory of specific events rather than being a routine response to death. By contrast, successive inhumation in caves and rock shelters took place within a much more fixed and enduring kind of place. Although caves were certainly active environments and would have had a significant effect on the bodies deposited there, their existence as comprehensible places would have pre-dated their use for burial. Neither were most caves used for successive inhumation significantly modified as part of the burial rite.

These different temporalities for the place of burial can be drawn out by comparing chambered long barrow practice with a cave where we do know there was episodic successive inhumation in the Neolithic: Totty Pot in Somerset. Schulting, Gardiner and colleagues (2010, 80–81) modelled the Neolithic funerary activity there as taking place in three different phases – in the Early, Middle and Late Neolithic. The best-dated of the Totty Pot successive inhumation events is the one in the Middle Neolithic, which may even be a single event occurring at some time between *3355* and *2930 BC*. The significant difference between Totty Pot and the chambered long mounds is that episodic successive inhumation continued there much later, with the final event taking place almost at the end of the Late Neolithic. The cave burials were also both less spatially constrained and more effectively removed from the world of the living than those in the chambered long mounds. It is likely that bodily decomposition in the darkness zone of Totty Pot was a quantifiably different process to that which took place in the relatively crowded environs of, for example, the mortuary structure of Wayland's Smithy I. Bodies in the cave would have been transformed slowly, becoming partially coated in tufa as they decomposed (Schulting et al. 2010, 80), and they may not have even been accessible to living people after they were deposited. By contrast, bodies in the Wayland's Smithy mortuary structure would have decomposed faster, being exposed to a wider range of agents. If Whittle and colleagues (2007, 114–117) are right in their assumption that successive inhumation there happened over a very short timescale, then the almost simultaneous presence of these bodies in a constricted space would have accelerated the decomposition process significantly (Simmons et al. 2010).

The successive inhumation burials from Middle and Late Neolithic caves generally show this tendency towards a relatively more protracted form of the rite. The different temporalities of the place of burial seem to have led to a different temporality for successive inhumation itself: Early Neolithic burial in tombs was relatively quick while cave burial, particularly in the Middle and Later Neolithic, was relatively slow. The active contribution of the cave environment in these cases seems to have been to extend the duration of the intermediary period and protect the body from both human and natural agents that might have accelerated the process. I believe that this relatively slow version of the rite explains why successive inhumation in caves lasted so much longer. The protracted nature of the intermediary period, and the physical traces it left in the cave deposits, would have acted as material indices of the rite, enabling the details of successive inhumation in these places to persist over much longer timescales. Successive inhumation in cairns was quick, episodic and tied to the remembrance of particular events and people. In caves, it was slow and occasional, part of an extended routine of death which persisted throughout the Neolithic, as shown, for example, by the range of radiocarbon results from Raschoille in Argyll (Figure 8.4).

Cave burial and society

The comparison between the temporalities of successive inhumation in caves and cairns leads naturally on to the wider question of the relationship between cave burial, cairn burial and the wider response to death in Neolithic society. It has been suggested previously that cave burials were unusual or aberrant rites, or that there was some social distinction between the people who were buried in chambered cairns and those who were buried in caves (Schulting 2007, 591). In areas such as South Wales and south-west England, where there are significant numbers of dated individuals from both chambered cairns and caves, this question has been approached using the stable isotope evidence for past diets (e.g. see Schulting and Richards 2002a, b; Schulting et al. 2010; 2013). The argument has been that if there were significant status differences between the people who were buried in caves and those who were buried in chambered cairns, with the assumption usually being made that those buried in the cairns were the 'elite', then this should be reflected in differential access to certain food sources. The most recent review (Schulting et al. 2013, 22) of the comparative data between caves and cairns for the south-west concluded that there was no evidence of any significant difference in diet between the two groups of people. This is not to say that there were no status differences between the two burial populations, simply that they were not reflected in the bone chemistry data.

The relationship between caves and chambered tombs as burial spaces has been considered several times in the literature. In their pioneering synthesis of the evidence for the Neolithic use of caves, Barnatt and Edmonds (2002, 114) suggested that there would not have been any conceptual difference between caves and chambered tombs in the Neolithic. They suggested that both would have had similar physical properties and that it was highly likely that Neolithic populations regarded underground spaces like this as having been constructed by earlier people, possibly semi-mythical ancestors. Schulting (2007, 588) considered it highly unlikely that cave burial predated the use of cairns. Drawing on the evidence for successive inhumation from both kinds of site, he was of the opinion that cave burial was inspired by existing chambered cairn burial rites, which might tend to support Barnatt and Edmonds' case that the two kinds of site were regarded in the same way in the Neolithic. However, my detailed review of the practice of successive inhumation clearly shows that there were important differences in the way caves and cairns were used for some kinds of burial. Dowd (2015, 110–111) has made the case that a similar distinction can be recognised in Ireland between the use and perception of caves and chambered tombs.

Despite the differences noted here for successive inhumation nationally, there are some sites where it is difficult to see any distinction between burials which took place in caves and those which took place in chambered tombs. At Hay Wood Cave, Somerset (Everton and Everton 1972; Schulting et al. 2013, 12–15 and see Chapter 6), at least ten individuals were successively buried within a narrow chamber and passage over a relatively short period of time in the Early Neolithic. It is difficult to see any significant difference between this practice and the successive inhumation of nineteen individuals in the chambers and entrance passages at Hazleton North chambered cairn, Gloucestershire. The date and duration of the Hazleton North burial rites were modelled by Meadows and colleagues (2007, 61), whose preferred interpretation was of a similar short burial duration in the Early Neolithic with, as at Hay Wood Cave, the possibility of two phases of activity within this duration.

The use of caves for other kinds of burial rite at the beginning of the Neolithic also show the commonalities between caves and other kinds of burial space. As noted here, secondary burial appears to have taken place in both chambered cairns and caves. This was also the case for the midden burials described in Chapter 5. Therefore, there is a case for arguing that some caves were conceived of as being equivalent to some chambered tombs but that this was not universal. At a very broad level, very early cave burials were more likely to look like examples of rites which also took place in other kinds of environment. Later cave burials, by contrast, were typically long and slow successive inhumations, and it may be that in these we can identify a practice which only took place in caves.

The origins of cave burial practices were reviewed in detail in Chapter 5. There, it was argued that, at the very least, cave burial was one of the earliest manifestations of a Neolithic way of life in each of the regions where it occurs, and it is possible that there were a significant number of cave burials which took place in the very late Mesolithic. Of course, there is a significant element of circular reasoning involved when we try to distinguish between Late Mesolithic and Early Neolithic versions of what seem to be similar kinds of rite. Almost nothing about the human remains in these caves allows us to definitively identify them as being hunter-gatherers or early farmers. It is probably best to refer to these cave burials simply chronologically as early fourth-millennium BC practices. However, when we look for broader parallels and try to understand where these people learnt about multi-stage burial rites, then we see that almost all the proposed models for these practices are Neolithic. The recently modelled dates for the Coldrum chambered tomb (Wysocki et al. 2013, 11–13), where there was clear evidence for a multi-stage rite incorporating an intermediary period, show that monument began to be used between *3980* and *3800 BC* (at two standard deviations). This monument was not itself the earliest manifestation of Neolithic activity in the Thames region. Therefore, although early fourth-millennium BC cave burials in Yorkshire and South Wales look to be early in their regional sequences, they were not demonstrably earlier than the earliest Neolithic activity in south-east England, or indeed than Neolithic activity in the Low Countries and northern France. Early cave burial rites, in their multifarious forms, seem to me to be an excellent example of the 'unpacked' Neolithic revolution. They were an aspect of Neolithic life which disseminated rapidly amongst transitional groups; they sometimes appropriated older places, such as middens; and they provide early evidence of contact and communication between hunting-gathering and farming groups. Recent Bayesian analysis of the earliest use of cereals in Britain (Griffiths 2016) shows a similar but opposite phenomenon as cereals were not adopted as rapidly as some other Neolithic things. For this reason, I think that Schulting and colleagues (2013, 22) are right to stress the essentially 'Neolithic' nature of early fourth-millennium BC cave burial, even though the dating evidence for some burials having taken place in what would chronologically be regarded as the Late Mesolithic is stronger than they allow.

In summary, one of the definable differences between the various cave and cairn burial rites identified in this book was that they had different temporalities. These temporalities were directly tied to the way that the burial space acted upon the dead. In all cases, there was an intermediary period involved in the rite, but the evidence suggests that there was a great deal of variability in how 'quick' or 'slow' burial transformations were. It also appears that 'slow' burial rites were more likely to persist over many centuries, while 'quick' ones were more likely to be episodic and confined to a few relatively intense bursts of activity. As I have discussed here, one of the major factors which the mourners would have been able to manipulate when choosing between different burial tempos is the kind of place that was chosen for the burial. The physical burial environment of caves would have varied enormously depending on the kind of cave chosen, the direction it faced, whereabouts within the system bodies were placed and how many bodies were deposited together.

Experiencing cave space

It has been suggested in other studies (Holderness et al. 2006, 82–85) that the direction in which a cave faced can be shown to be a significant factor in how it was used. Holderness and colleagues looked at the aspect directions of archaeological caves of all periods in both North Yorkshire and part of the Midlands. They established that there were possibly two slight trends, with preferential use of caves facing towards the east and west-south-west. However, these trends were not statistically significant in either case. Figure 8.5 shows the aspects of all of the cave burial sites analysed in this book. This demonstrates that even within the sub-set of caves used for burial during the Neolithic, there was a great deal of variability, with the only overall trend being the low numbers of east and west-facing caves chosen. This trend and the lack of consistency in this national data probably indicates that people did not chose caves with reference to distant cosmological events such as the rising or setting of the sun. The fact that the Neolithic data do not match the results from the study quoted here strongly suggests that if cave aspect was a factor in

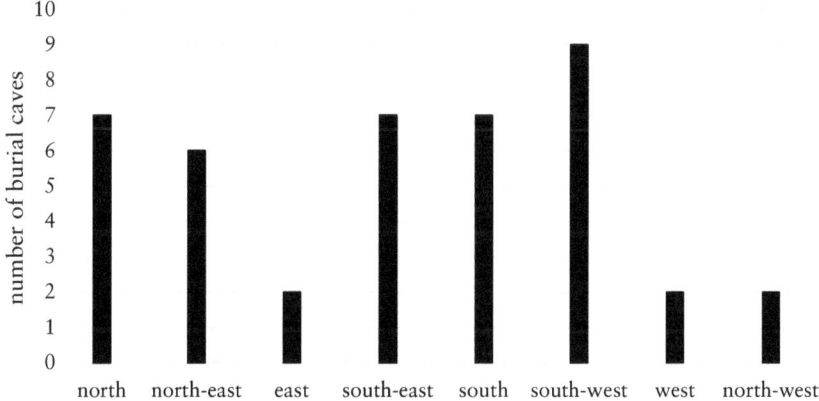

8.5 The direction of aspect from the mouth of forty-two of the forty-eight caves in the study (dolines and other vertical entrance caves have been excluded).

how the burial was experienced, this was specific to certain times and places. Certain rites in particular places and times may have required the aspect of the caves to be chosen for local, phenomenological, reasons. The details vary from region to region, but there was clearly a conceptual link between the cave as a place of burial and the wider social and environmental structure around the cave.

One example of this local phenomenon can be seen in the Yorkshire Pennines. Figure 8.6 shows how caves with Neolithic human remains in that region face in directions ranging from south-east to south-west. The only doline in this group, North End Pot, is also situated on a south-west-facing slope. The distribution map also shows that these sites are on the southern and western fringes of a wider group of usable caves. Therefore, it may be that their orientation is a result of a decision to prefer caves along the south-western edge of the Pennine massif. Figure 8.7 shows the view from the mouth of Kinsey Cave, giving a good indication of the way that these sites generally face away from the fells and towards the more open river valleys and more distant lowlands.

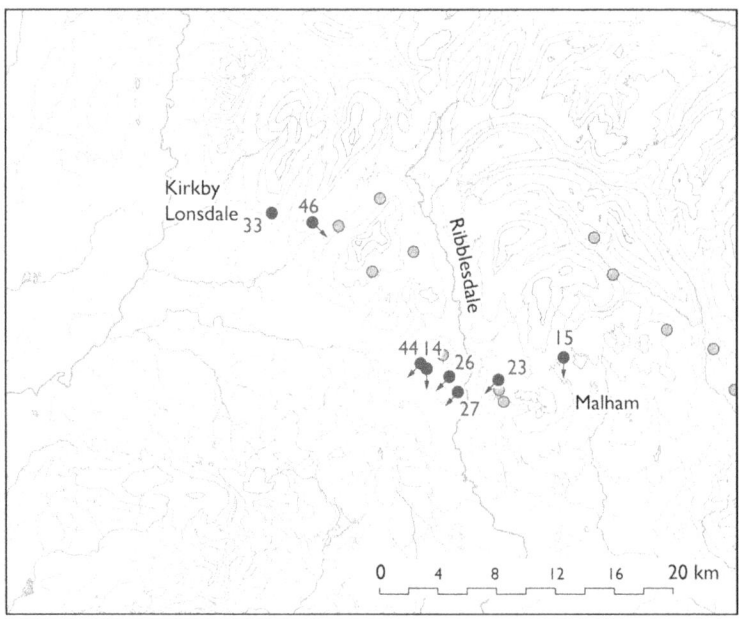

8.6 Aspect directions of caves in the Yorkshire Pennines with Neolithic human remains: 14 – Cave Ha 3; 15 – Chapel Cave; 23 – Jubilee Cave; 26 – Kinsey Cave; 27 – Lesser Kelco Cave; 33 – North End Pot; 44 – Sewell's Cave; 46 – Thaw Head Cave. The unlabelled grey dots are caves which contain human remains of other dates (data from Chamberlain 2014). Base mapping derived from Ordnance Survey data © Crown copyright/database right 2016. An Ordnance Survey/Edina supplied service.

8.7 View from the mouth of Kinsey Cave, facing south-west away from the high fells of Pen-y-Ghent and Ingleborough and across Giggleswick Common towards the Forest of Bowland.

In south-west England, there is another group of Neolithic burial caves which seem to have been chosen to provide a consistent aspect from the cave mouth. Those caves on the southern fringes of the Mendips which were chosen for burial often have a northerly aspect, facing towards the higher ground. Chelm's Combe, Flint Jack's Cave, Hay Wood Cave and the postulated original entrance to Totty Pot all face due north, and Picken's Hole faces north-east. Therefore, in both the Yorkshire and Somerset examples, cave aspect was clearly an important part of why the site was chosen for burial. However, the differences between the two regions show that it was local phenomenological considerations which influenced this choice. It also seems likely that the reasons behind the choice were different in each region too. The Yorkshire caves generally look outwards towards lower-lying land, while those in Somerset tend to face inwards towards the uplands.

At a smaller scale, there was also clearly a concern with the immediate environment within the cave. I have discussed here how differences in the cave environment might affect the temporality of the intermediary period. It is also likely that the enclosing effect of the cave architecture itself was reflected in the positions of burials within the caves. This is particularly the case with the successive inhumation burials discussed in Chapter 7 from the Middle and Late Neolithic. These burials are the remains of a practice which seems to have been occasional and protracted. In contrast to all the different burial rites which were used in the earlier part of the period, it was also one in which the living mourners had little input. The evidence from this period of human bone deep within underground systems suggests that bodies were being deliberately placed where the combined agency of their decomposition and the processes of cave sedimentation would provide an extended intermediary period which was physically removed from the world of the living. At sites such as Reindeer Cave, Inchnadamph; Totty Pot; North End Pot; and Orchid Cave, bodies were placed in such a way that most of the intermediary period would take place within the deeper parts of cave systems. It is not clear in any of these cases how far into the cave the initial place of deposition was, but it is clear that those carrying out the burial understood that by placing the bodies there they were ensuring that processes would move them deeper into the system. This Middle and Late Neolithic version of the successive inhumation practice in caves might be described in terms of a journey from the widest landscape perspective to the deepest part of the cave system. In the earlier Neolithic, there already seems to have been some distinctive elements to the successive inhumation rite in caves. As discussed here, it was comparatively slow and occasional when compared to similar rites in chambered cairns. At least a part of that difference arose from the differences in the physical environment between the two kinds of sites. By the Middle Neolithic, people who were choosing to carry out successive inhumation burial were deliberately drawing on the agency of decomposing bodies and deep cave spaces to create a strongly defined burial rite which appears to have been exclusive to caves. Metcalf and Huntington (1991, 99–100) described the intermediary period among the Toradja as one in which the decomposing body provided a physical index of the stage that the soul has reached on its journey to the underworld. Adopting this as an analogy, it is tempting to see this rite as a deliberate strategy to ensure that both the decomposing physical body and the voyaging soul (or at least the network of unravelling social obligations and bonds created by the death of the individual) moved in a controlled and appropriate way on their journey. In this case, movement in the intermediary period started in the social world of the living mourners but was largely the concern of the underground agency of caves and bodies.

Material worlds of death

I suggested towards the end of Chapter 4 that it is helpful to think of these kinds of material engagements as relational. Therefore, we would expect to see examples where other objects and sediments within the caves would become drawn into the entangled biographies of practice which contributed to the development and maintenance of these specific funeral rites. One example, which has been considered elsewhere (Leach 2008, 51), is the possible association between some burials and caves with active deposition of tufa. Alongside Thaw Head

Cave and Cave Ha 3, the two Yorkshire sites discussed by Leach (2008, 39–41), skeletal remains were associated with tufa deposits at Totty Pot, Flint Jack's Cave and Chelm's Combe in Somerset; at Nanna's Cave and Ogof-y-Benglog on Caldy Island; and at George Rock Shelter in the Vale of Glamorgan. Leach (2008, 51) suggested that the material engagement with tufa was important because its petrifying qualities acted to preserve the body, extend the duration of the decomposition and, to a certain extent, to mask or neutralise some of the smells and visual signs of decomposition. Therefore, we could suggest that the properties of tufa were being used to extend and slow down the intermediary period. Given that I have identified that one of the distinctive aspects of successive inhumations in caves is that they have a 'slow' tempo when compared to cairn burials of the same type, we could argue that caves with actively forming tufa were chosen to help create this 'slow' intermediary period. However, in two of the cases discussed here, Chelm's Combe and George Rock Shelter, the tufa seems to have been used in the intermediary period of a secondary burial rite. Therefore, I would argue that the importance of tufa provides a relational link between different funerary rites. Both successive inhumation and secondary burial associated with tufa share a distinctively different tempo to other Early Neolithic burials, which arises from the fact that they both take place in caves.

A similarly slow tempo for successive inhumation might be suggested at those cave and rock-shelter sites where human remains are associated with middens. At Raschoille, Carding Mill Bay 1 and An Corran, the available radiocarbon evidence shows that burial was episodic over an extended duration (see Chapter 5). There is some evidence from An Corran that the midden there was still being added to when the burials took place, although at all three sites, the middens seem to have been established before burial began and to be primarily Mesolithic structures. Unlike tufa, midden material was not a part of the cave environment which could actively contribute to creating a slow tempo. However, middens would have provided a material indication of both the extended occupation of the landscape and cave and of past food consumption. The Scottish sites may also have commemorated the seasonal nature of the exploitation of marine resources. These middens are unlikely to be the remains of food consumed as part of the funerary rites, and indeed the dietary stable isotope data from Carding Mill Bay 1 (Schulting and Richards 2002b, 155–157) suggests that people being buried at this site did not have a significant marine component to their diet. If the middens were being actively added to as part of the funerary activity associated with the intermediary period, this was probably a deliberate echo of earlier practices. It is also possible that older midden material was being brought to the rock shelters and deposited as part of the intermediary period rites. There is a possible correlation, therefore, with the terrestrial midden burial at Broken Cavern, Devon, where the soil micro-morphological evidence suggests that the midden material was moved into the cave (Collcutt in Roberts 1996, 203). These burials drew on the mnemonic properties of the middens to link the intermediary period with earlier kinds of seasonal food gathering and subsistence. It may be that the well-documented cultural shift away from marine resources at the beginning of the Neolithic (Richards and Schulting 2006) was embedded in an origin myth which linked shell-fish and the sea with ancestral beings or lineage founders, in which case, the desire to use and even create shell-middens as a part of the intermediary period in funerary rites may have reflected this belief. As the corpse decomposed, it would have shifted from the social world of the living, apparently strongly tied to the production and consumption of meat and grain, to the domain of the ancestors, connected to the sea and its resources.

At Scabba Wood Rock Shelter, South Yorkshire (see Chapter 7). there appears to be good evidence of the way that material culture was used during a particular kind of intermediary period. In common with other successive inhumation burials from the Middle Neolithic, the intermediary period at this site was separated from the active involvement of living people. The osteological evidence suggests, despite the open nature of the rock shelter, that the body was protected during at least the initial stages of the intermediary period. This sense of separation can also be seen in the pattern of deposition of worked stone at the site. I would suggest that the distinction between the distributions of worked stone and human bone fragments at Scabba Wood Rock Shelter is the result of objects being deposited at the site by visiting mourners during the intermediary period, but that these mourners had no access to the area where the bodily decomposition was taking place. As noted here, this tendency to separate the social aspects of the intermediary period continued into the Middle and Late Neolithic. The decomposition of the body, although presumably still a theoretical marker of the temporality of the intermediary period, increasingly took place in locations which were either secluded from view or inaccessible to the living mourners.

Cave burial in an inhabited environment

These studies of the physical characteristics of burial caves and their relationship to the artefacts and environment within the cave show the importance of understanding the relational links between caves as places of burial and the wider inhabited environment. Burial caves would have been only one aspect of the world which Neolithic people encountered.

To get a clear understanding of the way that caves would have acted in structuring funerary rites, it is helpful to examine some specific examples of how burial caves connected to the archaeology of both the immediate environment and the wider region. The first of these case studies concerns the early fourth-millennium BC site at George Rock Shelter, Goldsland Wood. As described in Chapter 5, the burials at this site seem to have been examples of secondary burials, with George Rock Shelter having been the location for the intermediary period before the bulk of the bones were moved to a final burial site elsewhere.

Goldsland Wood lies on a limestone ridge at the east end of the Vale of Glamorgan in south-east Wales (Figure 8.8). George Rock Shelter is particularly suitable for this kind of landscape level study. First, the fact that secondary burial took place implies that the network of relationships around the funerary rite connected more than one place in this landscape. In addition, the Vale of Glamorgan has a well-documented and relatively rich Early Neolithic archaeological record, which enables us to reconstruct some sense of the density of inhabitation around the site (Figure 8.9). If we explore the network of relationships around the secondary burial rite, there are a number of possible places where the disarticulated remains may have been taken after the intermediary period. There are several early Neolithic chambered tombs within a few

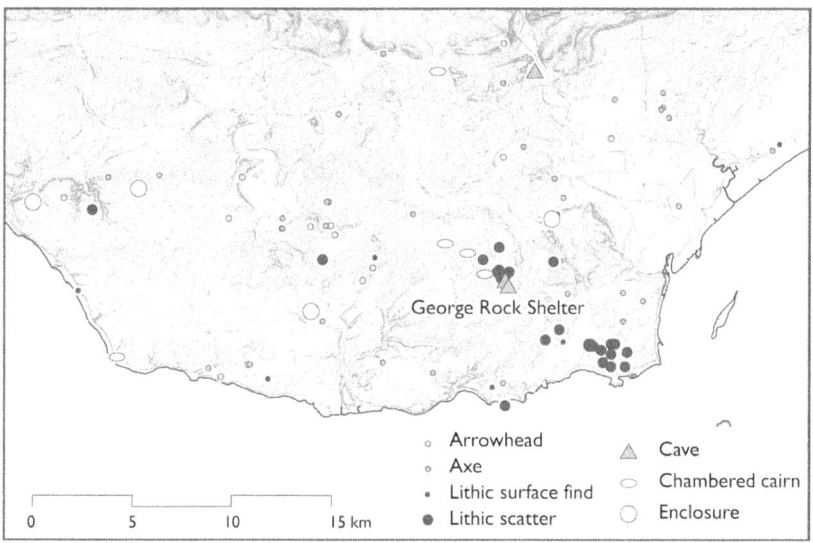

8.8 The location of George Rock Shelter and of documented Early Neolithic archaeology on the Vale of Glamorgan. Archaeological data from the Glamorgan-Gwent Archaeological Trust Historic Environment Register with some additions. Base mapping contours at 10-metre intervals derived from OS data © Crown Copyright and Database Right (2018). Ordnance Survey (Digimap Licence).

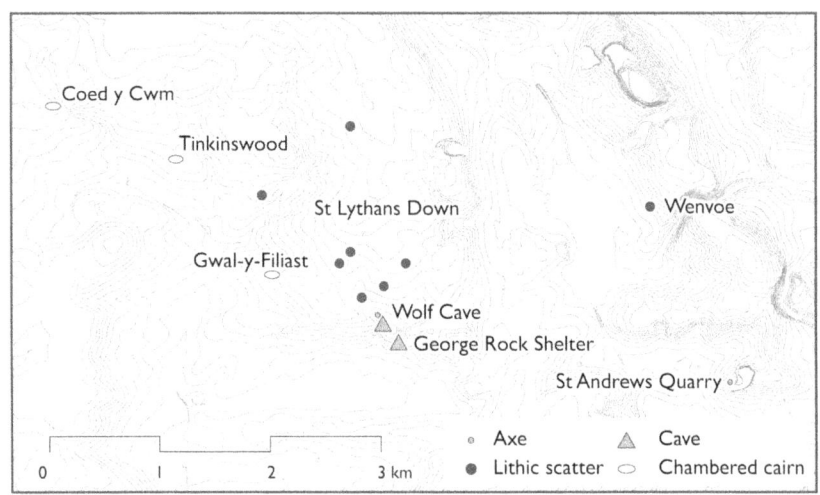

8.9 The location of George Rock Shelter and the Early Neolithic archaeology in its immediate environs. Archaeological data from the Glamorgan-Gwent Archaeological Trust Historic Environment Register with some additions. Base mapping contours at 5-metre intervals derived from OS data © Crown Copyright and Database Right (2018). Ordnance Survey (Digimap Licence).

kilometres of the site. The nearest of these is Gwal-y-Filiast, St Lythans, which is around 1 km to the west of George Rock Shelter, at the other end of the limestone ridge. This monument was excavated at some time before 1875, and J.W. Lukis recorded both human bone and pottery from the chamber. Several years ago, a leaf-shaped arrowhead was recovered as a surface find from the chamber area (Evans and Lewis 2003, 6–7; Cedric Mumford, personal communication). The chambered tomb at Tinkinswood, St Nicholas, also contained pottery and what appears to have been an extensive collection of human remains (Reynolds 2014, 176–178; Ward 1916, 243–244). Slightly further to the west, three fallen blocks at Coed y Cwm have also been identified as another possible chambered tomb, in this case associated with a single find of a polished flint axe (Evans and Lewis 2003, 6). Any or all of these sites may have been the final location for the human remains which spent their intermediary period in George Rock Shelter.

There is evidence from lithic scatters on St Lythans Down for intensive occupation in the immediate vicinity of the site. The multi-period nature of these scatters may reflect the multi-period nature of the lithic assemblage at George Rock Shelter itself, which clearly remained an important place within the landscape long after the dated period of burial activity. Another of the caves within the Goldsland Wood complex, Wolf Cave, has produced Early Neolithic pottery and lithics alongside highly fragmented human remains. In this case, radiocarbon results show that some of these human remains were Early Bronze Age and others were early medieval. Wolf Cave is therefore probably best regarded as an example of Neolithic inhabitation evidence associated with caves. There was also a surface find of a polished flint axe from a ploughed field immediately to the north of Wolf Cave (David Randolph, personal communication), which is further evidence for a very high density of Early Neolithic occupation close to the cave complex.

There are similar single surface finds of lithics over the wider Vale of Glamorgan (see Figure 8.8), particularly of polished axes and leaf-shaped arrowheads. There is an element of collection bias in this distribution, as axes and leaf-shaped arrowheads are both easily recognisable artefact types. However, their recovery probably indicates, alongside high densities of lithic scatter evidence in those areas where fieldwalking and developer-funded archaeology has taken place, that Early Neolithic inhabitation in the Vale of Glamorgan was equally intensive over the whole region. There is one further example of a cave associated with inhabitation evidence. A small number of Neolithic flint artefacts were found in Lesser Garth Cave, north of Radyr (Madgwick et al. 2016, 207), alongside a much larger collection of later prehistoric and medieval artefacts and human remains, paralleling the discoveries at Wolf Cave. In addition to the monuments immediately adjacent to George Rock Shelter noted here, there is a ruined chambered tomb overlooking the Ely at Cae'r Arfau with another site at the west end of the Vale at Cae'r Eglwys near Nash Point (Evans and Lewis 2003). There is also now considerable evidence for earlier Neolithic enclosures in the region. There are probable causewayed enclosures at Norton, Ogmore-by-Sea, Flemingston and Corntown (Davis and Sharples 2017, 19–21). However, the site at Caerau, Ely, provides the best evidence for an earlier Neolithic enclosure. The multiple circuits of interrupted ditches at this site have produced a relatively small lithic assemblage and a substantial group of Early Neolithic pottery sherds (Davis and Sharples 2017, 8–9). Radiocarbon dating of the Caerau enclosure (Davis and Sharples 2017, 12–13) shows that the site was being used between 3600 and 3400 BC and supports the suggestion (Whittle et al. 2011, 548–549) that causewayed enclosures in South Wales were not constructed until at least the thirty-seventh century BC, considerably later than the suggested date for George Rock Shelter.

Cave burial in the Vale of Glamorgan appears to have been both a relatively rare practice and to belong very early in the local Neolithic sequence. Cave sites themselves are not very common in this region, but it is noticeable that of four excavated examples, only George Rock Shelter has produced dated evidence for Neolithic burial. However, the secondary burials at this site took place within what seems to have been a densely settled landscape. They also provide excellent examples of relational links between the cave burial site and other parts of the landscape. At the end of the intermediary period, the bones from George Rock Shelter were almost certainly moved to one of the local chambered tombs. Similar Early Neolithic material culture, particularly pottery and leaf-shaped arrowheads, were used at both kinds of site. Interestingly, there is also good evidence of the ways that this material culture acted as circulating references to provide continuity and temporal connections. George Rock Shelter and the other caves in the region continued to be important places in the landscape; Middle and Later Neolithic material culture was deposited there long after the cave burial rite had apparently ceased.

I have attempted a similar landscape case study for the group of four burial caves which are found along the edge of Giggleswick Scar in North Yorkshire. Some of the sites here were also in use very early in the fourth millennium, but there are interesting differences in both the surrounding archaeology and the wider geographical setting. As discussed here, the caves are found in the south-west facing escarpment of Giggleswick Scar, from Sewell's Cave at the northern end to Lesser Kelco Cave in the south (Figure 8.10). They include examples of both successive inhumation burial and the isolated burial of crania and have date ranges which cover almost the whole of the Early Neolithic. Archaeological evidence for the wider region suggests that there were relatively dense levels of Neolithic inhabitation around these sites. There are obvious collection biases when compared to the more intensively farmed Vale of Glamorgan, but a long history of active local archaeology has provided a relatively complete record.

As noted here, Giggleswick Scar is on the south-western edge of this part of the Yorkshire Pennines. There is evidence from the caves along Attermire Scar, about 3 kilometres to the east, of Neolithic activity and occupation. This cluster of sites – Bat Cave, Albert Cave, Attermire Cave and Horseshoe Cave (see Figure 8.11) – all have reports of Peterborough Ware pottery or lithics. This is often in small quantities amongst material of other periods, and none of the caves have produced dated Neolithic human remains (Dearne and Lord 1998; Jackson 1953). Other evidence of contemporary inhabitation in the landscape nearby includes a burnt mound at Attermire Scar, which is probably either Late Neolithic or Early Bronze Age. There is also a cup-marked stone close to the Attermire Scar burnt mound, and another was recorded at Lower Winskill (Northern Archaeological Associates 2002). A stone axe was discovered during nineteenth-century drainage works at Crow Nest Farm, around 500 metres to the north-west of Sewell's Cave (Compton 1892, 79).

A wider area of the local landscape provides a more complete view of the kinds of activity which were taking place in the landscape

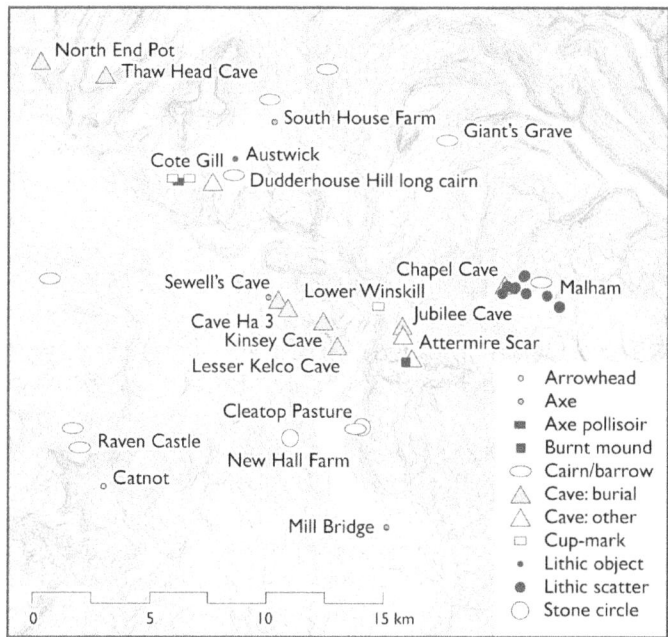

8.10 The location of Giggleswick Scar and of documented Neolithic archaeology in the surrounding landscape. Archaeological data from North Yorkshire Historic Environment Register and Yorkshire Dales National Park Historic Environment Register with some additions. Base mapping contours at 10-metre intervals derived from OS data © Crown Copyright and Database Right (2018). Ordnance Survey (Digimap Licence).

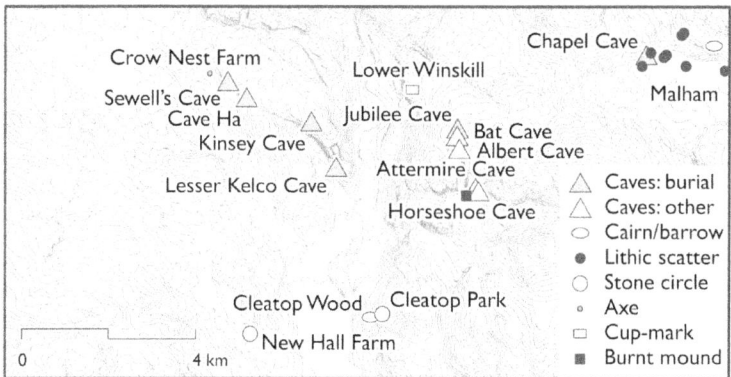

8.11 The location of Giggleswick Scar and the possible Neolithic archaeology in its immediate environs. Archaeological data from North Yorkshire Historic Environment Register and Yorkshire Dales National Park Historic Environment Register with some additions. Base mapping contours at 10-metre intervals derived from OS data © Crown Copyright and Database Right (2018). Ordnance Survey (Digimap Licence).

around Giggleswick Scar (Figure 8.10). Where appropriate survey has taken place, lithic scatter evidence is very dense. The cluster of sites (Williams et al. 1987, 379–381) around Malham Tarn provides evidence of occupation in the same landscape as the probable successive inhumation burials at Chapel Cave. These sites are all multi-period lithic scatters with a substantial Mesolithic component but with significant amounts of both Early and Later Neolithic material. There is also a surface find of a polished flint axe from Great Close within this cluster. Single surface finds of artefacts from the rest of the landscape show that this level of occupation was probably not unusual. A leaf-shaped arrowhead was discovered at Catnot (King 1970, Figure 25), a Late Neolithic blade from Austwick and other stone axes have been reported from South House (Gilks and Lord 1985) and Mill Bridge. There are also probable axe-polishing grooves on a boulder at Cote Gill (Yorkshire Dales National Park HER records).

The greatest contrast with the Vale of Glamorgan landscape is in the sheer number of caves in this area which were used for burial in the Neolithic. In addition to the four sites on Giggleswick Scar and Chapel Cave, there are also the two more northerly sites of Thaw Head Cave and North End Pot, dating to the beginning and end of the Neolithic, respectively. This may be connected with the generally wider use of caves in the region during the Neolithic. The other contrast with the South Welsh evidence is the rock-art sites within the local landscape. As well as the examples at Attermire Scar and Lower Winskill, there are cup-marked stones at Cote Gill (Schofield and Vannan 2014, 10–11). There is also some evidence for Neolithic monuments in this landscape. To the south, Late Neolithic dates have been suggested for stone circles at New Hall Farm and Cleatop Pasture (Yorkshire Dales National Park HER records). Of the cairns and barrows recorded in the two regional Historic Environment Registers, Raven Castle is listed as a possible chambered tomb. Dudderhouse Hill long cairn appears to have had a forecourt and some indications of former chambers in the denuded central area. The excavated round mound at Giant's Graves (Bennett 1938) had evidence of internal chambers and was considered by the excavator to be Late Neolithic.

The most striking aspects of the way that the burial caves on Giggleswick Scar relate to their wider environs is the way that all four sites lie on the edge of the documented areas of inhabitation. The burial caves face away from this inhabited landscape so that there is a clear sense of burial caves as a separate and different kind of space. This is despite the fact that there is much more evidence for the centrality of caves to all aspects of the local Neolithic. There was also apparently a longer timescale to the development of cave burial practices in this area. In these cases, remembering that bodies, caves and the actions of the living would have all been the kinds of event which constituted the temporality of this landscape, we may be able to see how the persistent presence of a visible cave burial rite and visible landscape use acted as indices of an old environment.

In both of these regions, we can see how the network of relationships around burial caves was an integral part of how the local Neolithic was experienced. Bodies acted, moving and being moved from one set of relationships and understandings as living people to another as active corpses. Caves acted as circulating references, as indices of change and temporality. Material culture and the people who used it acted to link together caves and the wider landscape. It is notable that the physical experience of the wider landscape in these two regions would have been radically different. In the Vale of Glamorgan, once immediately away from the funerary cave, people would have been moving through areas of deep and fertile soils and over relatively gentle gradients. In the Yorkshire Dales, the landscape between the caves is largely of thin soils and exposed limestone pavements, and movement from one outcrop to another would have involved much more challenging terrain. However, it is noticeable that, despite these differences, the immediate approach to all the caves in both study areas is very similar. All the sites would have been accessible only after steep climbs over difficult limestone screes. It may be that this similarity of external experience is an important part of the way that the cave sites acted. In an analogous manner to the central Philippine ethnographic examples described in Chapter 3, the journey that artefacts and bodies undertook up to the caves would have provided an important and dramatic commemorative event.

Conclusions

Hopefully, over the course of this book, I have demonstrated both the range and importance of Neolithic funerary practices from caves in Britain. Regardless of the detail of my interpretation of these rites, the volume of evidence from all the regions of Britain with suitable caves shows that cave burial must have formed a significant strand in Neolithic funerary practice. Working from the standpoint that most of the human remains were the result of deliberate multi-stage collective burials (in common with other Neolithic human remains), then I have argued that the most effective way to understand how the intermediary period worked is to look for relational material evidence. This evidence shows how the social implications of death for the living, the biological agency of decomposition in the dead body and the environmental agency of caves as particular kinds of place worked together to constitute various different versions of the intermediary period.

In the earlier part of the Neolithic, a range of different fragmentation techniques were appropriate during the intermediary period. It is noticeable that it is at this time that we have evidence for at least two different secondary burial rites and possibly for other techniques which involve the active intervention of the living. The distinguishing thing about multi-stage funerary practice in the Early Neolithic is that the remains continue to have elements of this social agency right up to and beyond the point where they have been fragmented. They are accessible to the living, and the living appear to be interacting with them. After about 3300 BC, the shift to a sacred agency seems to have taken place much more quickly. The dead moved rapidly into a state where, however much sacred agency they had to act on other people, places and objects, it was not appropriate for social agents to act upon them.

I would argue that the particular cave burial rites identified in the Middle and Late Neolithic developed specifically because the intermediary period was taking place in caves. In the Early Neolithic, burials in caves appear to be examples of a range of different burial practices which could possibly occur at either cave or non-cave locations. The secondary burial of crania, successive inhumation and secondary burial can all be paralleled from other landscape and monumental locations. However, as I have argued here, differences between the detail of the intermediary period can be detected even in these Early Neolithic examples. In particular, the temporality of cave burial seems to be significant different. By the Middle Neolithic, the interaction of the physical agency of the cave, the social agency of the living and the sacred agency of the dead during these varied rites had caused them to coalesce into a rite which emphasised the special nature of caves as slow transformative spaces. By this date, the journey from the social network of the living to the sacred network of the dead was something that was bound up with the transformative properties of the caves themselves.

Caves and cave landscapes would always have acted within the funerary rites. They would have been important and distinctive spaces. However, by studying the details of these funerary rites, it has been possible to demonstrate how the distinctiveness of caves was created. People, bodies and caves were all part of this process. Caves did not derive their ability to act solely from their physical distinctiveness, nor were they passive spaces which were imbued with meaning by human actors. They became distinctive because they were repeatedly used for funerary rites. Throughout the Neolithic, we can see this process intensifying as caves became more active in and more central to the funerary process.

Appendix 1: Radiocarbon-dated human remains from British caves between 4000 and 2400 BC

Calibrated date ranges in this table were calculated using OxCal 4.3 and the IntCal 13 calibration curve (Bronk Ramsey 2009: Reimer et al. 2013).

Site name	No.	NGR	Lab no.	Bone sampled	Ident. no. and context	Date BP	Error	Cal. date ranges BC (2 Σ)	δ13 C (‰)	Reference
An Corran Rock Shelter	1	NG 4915 6848	OxA-13549	left navicular tarsal	AC/HB0627, context 31	4650	55	3635–3560: 3540–3335: 3205–3195	−19.4	Bronk Ramsay et al. 2009, 330
			OxA-13552	cervical vertebra	AC/HB0458, context 36	4535	50	3490–3470: 3375–3085: 3050–3035	−19.9	Bronk Ramsay et al. 2009, 330
			AA-27744	metatarsal III	AC/HB0628, context 31	4405	65	3335–3210: 3195–3150: 3140–2900	−20.2	Saville 1998
			OxA-13550	lumbar vertebra	AC/HB0632, context 31	4360	55	3330–3215: 3180–3155: 3125–2880	−20.5	Bronk Ramsay et al. 2009, 330
			AA-27743	ulna	AC/HB0270, context 36	3885	65	2565–2525: 2500–2195: 2175–2145	−24.0	Saville 1998
Ash Tree Cave	2	SK 5148 7615	OxA-4446	left tibia	burial 1/Sh. 10c, beneath scree	3730	90	2460–2415: 2410–1910	−21.1	Hedges et al. 1996, 397
Ash Tree Rock Shelter	3	SK 5145 7620	OxA-27992	right upper incisor	unstrat. surface find	4669	31	3620–3610: 3525–3365	−20.8	Dinnis et al. 2014
Ashberry Windypit 1	4	SE 5709 8501	OxA-13003	mandible	AS 165, chamber D	3873	28	2465–2280: 2250–2230		Leach 2006
			OxA-14439	humerus	AS AP 1, chamber D	3773	30	2295–2130: 2085–2055		Leach 2006
Backwell Cave	5	ST 4924 6801	BM-3099	vertebra	M6.12.52/54, charcoal-rich deposit	4510	40	3365–3090	−21.8	Ambers and Bowman 2003, 532

Bob's Cave	6	SX 5739 5124	OxA-4983	femur	silty cave earth deposit	5035	70	3970–3690: 3680–3665	−20.3	Hedges et al. 1998
Bower Farm	7	SK 0303 1954	OxA-16866	rib	(3) context 1024	4725	35	3635–3550: 3545–3495: 3465–3375	−20.7	Meiklejohn et al. 2011; Blockley 2006
			OxA-16864	cranium	cave entrance (unstrat.)	4724	33	3635–3550: 3540–3495: 3460–3375	−21.7	Meiklejohn et al. 2011; Blockley 2006
Blue John Cavern	8	SK 1319 8320	GU-21803	r. adult tibia midshaft	boulder choke	4125	40	2875–2575	−21.5	Nixon 2011
Broken Cavern	9	SX 8150 6748	OxA-3206	tooth	BRKFA-513, midden layer	4885	90	3945–3855: 3845–3825: 3835–3510: 3425–3380	−21.0	Schulting and Richards 2002a; Hedges et al. 1996
Cae Gronw Cave	10	SJ 0152 7108	OxA-5731	radius	86.32H, context 1	3955	60	2625–2280: 2250–2230	−20.9	Aldhouse-Green et al. 1996
Carding Mill Bay 1	11	NM 4874 2935	OxA-7664	human bone	C-XV:1	4830	45	3705–3520	−20.9	Bronk Ramsey et al. 2000; Hedges et al. 1993
			OxA-7663	human bone	C-XIV:1	4800	50	3695–3680: 3665–3505: 3430–3380	−21.5	Bronk Ramsey et al. 2000; Hedges et al. 1993
			OxA-7665	human bone	C-VII	4690	40	3630–3580: 3535–3365	−21.4	Bronk Ramsey et al. 2000, 461; Hedges et al. 1993
			OxA-7890	human bone	C-XXIII	4330	60	3315–3235 3110–2865 2805–2775	−22.0	Bronk Ramsey et al. 2000; Hedges et al. 1993

Site name	No.	NGR	Lab no.	Bone sampled	Ident. no. and context	Date BP	Error	Cal. date ranges BC (2 Σ)	$\delta 13$ C (‰)	Reference
Cathole Cave	12	SS 5377 9002	OxA-16605	cranium	NHM M.114	4675	39	3630–3530: 3595–3360		Schulting pers. comm.
Cattedown Cave	13	SX 493 536	OxA-15256	left upper 2nd molar	1308, north chamber breccia	4990	32	3940–3870: 3810–3690: 3680–3665	−20.3	Higham et al. 2007
Cave Ha 3	14	SD 7890 6624	OxA-13539	tibia	1 of 4, tufa deposit	4808	32	3655–3620: 3610–3520	−21.0	Griffiths 2011
			OxA-14266	juvenile mandible	4 of 4, tufa deposit	4595	40	3515–3395: 3385–3320: 3275–3265: 3240–3110	−22.0	Griffiths 2011
Chapel Cave	15	SD 88100 67200	OxA-V-2138–07	femur	MCC02, context 8, level 23–24	4805	45	3695–3680: 3665–3510: 3425–3380	−21.6	Blockley 2006
			OxA-V-2138–09	phalanx	MCC05, context 9, level 34	4715	40	3515–3395: 3385–3320: 3275–3265: 3240–3110	−21.6	Blockley 2006
Charterhouse Warren Farm Swallett	16	ST 4936 5457	OxA-1559	scapula	USF-393, horizon 2	3790	60	2460–2415: 2410–2115: 2100–2035		Hedges et al. 1989; Levitan and Smart 1989
			OxA-1560	neonate, femur	USF-394, horizon 4	3760	60	2440–2420: 2405–2375: 2350–2010: 2000–1975		Hedges et al. 1989; Levitan and Smart 1989

Chelm's Combe	17	ST 4634 5447	BM-2974	long bone	not known	4680	45	3630–3580 3535–3360	−22.1	Ambers and Bowman 2003
Darfar Crag Cave	18	SK 0975 5591	OxA-V-2137-51	vertebra	CnCo05	4914	33	3770–3640	−20.9	Blockley 2006
			OxA-V-2137-50	ulna	CnCo03	4669	33	3625–3605: 3525–3365	−21.5	Blockley 2006
Flint Jack's Cave	19	ST 4632 5381	BM-2839	right femur	1 of 4 inhumations	4430	80	3345–2915	−23.8	Ambers and Bowman 2003
Foxhole Cave, Paviland	20	SS 4385 8602	OxA-8315	phalange	FX32, context 2	4940	45	3895–3880: 3800–3640	−20.3	Schulting and Richards 2002a
			OxA-8318	phalange	FX177, context 3	4840	45	3710–3620: 3610–3520	−20.3	Schulting and Richards 2002a
			OxA-8317	tooth	FX59, context 1	4625	40	3620–3610: 3525–3335: 3210–3195	−20.6	Schulting and Richards 2002a
George Rock Shelter, Goldsland	21	ST 1121 7151	OxA-X-2424-44	1st upper right incisor	G385: individual 8, context 1004	5083	38	3965–3790	−20.0	Bronk Ramsey et al. 2015, 179
			OxA-20968	phalange	G1326, context 1002/1007	4929	33	3775–3645	−21.5	Higham et al. 2011

Site name	No.	NGR	Lab no.	Bone sampled	Ident. no. and context	Date BP	Error	Cal. date ranges BC (2 Σ)	δ13 C (‰)	Reference
Hay Wood Cave	22	ST 3398 5824	OxA-5844	cervical vert.	2Z10 AX/77/94	4860	65	3795–3515; 3400–3380	−20.8	Hedges et al. 1997
			OxA-19905	cranium	cranium I	4740	34	3640–3495; 3435–3375	−20.0	Schulting et al. 2013
			OxA-19904	cranium	cranium I	4742	31	3640–3500; 3430–3380	−20.1	Schulting et al. 2013
			OxA-19768	cranium	cranium II	4968	30	3890–3885; 3800–3660	−20.7	Schulting et al. 2013
			OxA-19906	cranium	cranium III	4786	32	3645–3515	−20.4	Schulting et al. 2013
			OxA-19907	cranium	cranium IV	4762	31	3640–3510; 3425–3380	−19.7	Schulting et al. 2013
			OxA-19908	cranium	cranium V	4770	45	3650–3500; 3435–3375	−20.2	Schulting et al. 2013
			OxA-19916	3rd max. molar	cranium V	4781	32	3645–3515; 3395–3385	−20.1	Schulting et al. 2013
			OxA-19882	3rd max. molar	cranium V	4748	31	3640–3505; 3430–3380	−20.3	Schulting et al. 2013
			OxA-19909	cranium	cranium VI	4723	32	3635–3555; 3540–3495; 3460–3375	−20.3	Schulting et al. 2013
			OxA-19917	cranium	cranium VII	4773	30	3645–3515; 3400–3385	−20.2	Schulting et al. 2013
			OxA-19910	cranium	cranium VII	4776	33	3645–3515; 3400–3380	−20.2	Schulting et al. 2013

Site	#	Grid ref	Lab code	Element	Context	BP	±	cal BC	δ13C	Reference
			OxA-19911	cranium	cranium VIII	4674	32	3625–3605; 3525–3365	−20.8	Schulting et al. 2013
			OxA-19912	cranium	cranium IX	4758	33	3640–3510; 3425–3380	−20.5	Schulting et al. 2013
			OxA-19881	cranium	cranium IX	4730	33	3635–3550; 3545–3495; 3455–3375	−20.4	Schulting et al. 2013
			OxA-19913	bone	burial 1	4851	31	3705–3630; 3580–3570; 3565–3535	−20.7	Schulting et al. 2013
			OxA-19914	tibia	adolescent	5052	32	3955–3770	−20.8	Schulting et al. 2013
			OxA-19915	lower canine	adolescent	5036	32	3950–3760; 3740–3730; 3725–3710	−20.5	Schulting et al. 2013
Jubilee Cave	23	SD 8376 6551	OxA-14262	tibia	individual 1	4836	31	3695–3625; 3590–3525	−20.9	Griffiths 2011
Kent's Cavern	24	SX 9344 6416	OS-36644	mandible and tooth	A5885, black mould	5020	45	3950–3705	−18.2	Meiklejohn et al. 2011
King Arthur's Cave	25	SO 5458 1558	OxA-5863	phalange	unstrat.	4670	60	3635–3550; 3540–3350		Chamberlain 2014
Kinsey Cave	26	SD 8040 6572	OxA-14799	mandible	scree and colluvium	5074	36	3965–3790		Lord et al. 2007
			OxA-15791	r. tibia	F005	5086	35	3965–3795	−20.5	Griffiths 2011
			SUERC-10518	l. tibia	F004	4820	40	3695–3675; 3670–3520	−21.4	Griffiths 2011
			OxA-15790	l. patella	F227	4472	33	3345–3080; 3070–3025	−20.8	Griffiths 2011

Site name	No.	NGR	Lab no.	Bone sampled	Ident. no. and context	Date BP	Error	Cal. date ranges BC (2 Σ)	δ13 C (‰)	Reference
Lesser Kelco Cave	27	SD 8098 6467	OxA-13538	cranium	individual 1, cave earth layer	4801	31	3650–3620: 3610–3520	−21.4	Griffiths 2011
Little Hoyle Cave	28	SS 1118 9997	OxA-3304	mandible	1983.2376/2	4930	80	3950–3630: 3580–3535	−21.2	Schulting and Richards 2002a
			OxA-3306	mandible	1983.2435/9	4880	90	3945–3855: 3820–3500: 3430–3380	−20.4	Schulting and Richards 2002a
			OxA-3305	mandible	1983.2376/11	4750	75	3655–3365	−19.9	Schulting and Richards 2002a
			OxA-3303	mandible	1983.2375/5	4660	80	3640–3320: 3275–3265: 3235–3170: 3165–3115	−19.4	Schulting and Richards 2002a
Little Orme Quarry	29	SH 8176 8248	Beta-87306	femur	individual 1, fissure fill	4720	50	3635–3485: 3475–3370		Gregory et al. 2000
Markland Grips	30	SK 510 751	OxA-4447	right mandible	individual 1	4760	90	3710–3360	−21.1	Hedges et al. 1996
			OxA-4448	right mandible	individual 2	4740	90	3705–3350	−21.6	Hedges et al. 1996
Mother Grundy's Parlour	31	SK 5358 7426	OxA-4442	right molar	E+FII,I,66, unstrat.	3720	80	2430–2420: 2405–2380: 2350–1895	−21.9	Hedges et al. 1996

Site	#	Grid	Lab code	Element	Context	BP	±	cal BC (95.4%)	δ13C	Reference
Nanna's Cave	32	SS 1458 9698	OxA-7739	femur	91.9H/4, disturbed	4560	45	3500–3460: 3380–3260: 3255–3095	−21.1	Bronk Ramsey et al. 2000
			OxA-7740	patella	63.355/61.1, NC2, disturbed	4520	45	3370–3085: 3050–3035	−21.2	Bronk Ramsey et al. 2000
North End Pot	33	SD 6830 7653	OxA-14265	mandible	NE19, lower shaft fill	4176	31	2890–2830: 2820–2660: 2650–2635		Leach 2006
Ogof-y-Benglog	34	SS 1470 9688	OxA-7743	vertebra	88.71H/2, unknown	4660	45	3630–3590: 3530–3355		Bronk Ramsey et al. 2000
Ogof Colomendy	35	SJ 2020 6277	SUERC-66486	molar	disturbed? layer 44	4408	33	3315–3235: 3110–2915		Ebbs 2017
Orchid Cave, Maeshafn	36	SJ 2002 6062	OxA-3817	pelvic frag.	92.23H, chamber fill	4170	100	3010–2980: 2940–2470		Aldhouse-Green et. al. 1996
Ossom's Crag Cave	37	SK 0958 5576	OxA-630	right ulna	O.VIII.3	4860	80	3915–3875: 3805–3495: 3435–3375		Gowlett et al. 1986
Picken's Hole	38	ST 3969 5500	OxA-5865	premolar	layer 3	4800	55	3695–3500: 3430–3375	−20.7	Hedges et al. 1997
Pontnewydd	39	SJ 0152 7102	OxA-5820	metatarsal	PN14, area A	4495	70	3370–3005: 2990–2930		Aldhouse-Green et. al. 1996

Site name	No.	NGR	Lab no.	Bone sampled	Ident. no. and context	Date BP	Error	Cal. date ranges BC (2 Σ)	$\delta 13$ C (‰)	Reference
Raschoille Cave, Oban	40	NM 8547 2888	OxA-8432	juvenile humerus	upper debris deposit	4980	50	3945–3855; 3825–3650	−20.4	Bonsall 2000
			OxA-8431	juvenile femur	upper debris deposit	4930	50	3910–3880; 3805–3635	−20.6	Bonsall 2000
			OxA-8433	adult humerus	upper debris deposit	4920	50	3895–3880; 3800–3635	−20.2	Bonsall 2000
			OxA-8441	adult humerus	upper debris deposit	4900	45	3785–3635	−21.2	Bonsall 2000
			OxA-8442	adult humerus	upper debris deposit	4890	45	3780–3630; 3555–3540	−21.0	Bonsall 2000
			OxA-8404	adult humerus	upper debris deposit	4850	70	3790–3505; 3430–3380	−21.6	Bonsall 2000
			OxA-8443	adult humerus	upper debris deposit	4825	55	3710–3510; 3425–3380	−20.4	Bonsall 2000
			OxA-8434	juvenile femur	upper debris deposit	4720	50	3635–3485; 3475–3370	−20.2	Bonsall 2000
			OxA-8444	adult humerus	upper debris deposit	4715	45	3635–3550; 3545–3485; 3475–3370	−21.1	Bonsall 2000
			OxA-8435	adult humerus	upper debris deposit	4680	50	3635–3565; 3540–3360	−22.5	Bonsall 2000
			OxA-8400	adult rib	lower debris deposit	4640	65	3635–3550; 3545–3320; 3235–3170; 3165–3115	−20.3	Bonsall 2000

			OxA-8399	vertebra	lower debris deposit	4630	65	3635–3560: 3540–3315: 3275–3265: 3240–3110	−21.4	Bonsall 2000
			OxA-8401	juvenile femur	upper debris deposit	4565	65	3520–3395: 3385–3085: 3060–3030	−21.1	Bonsall 2000
			OxA-8537	juvenile humerus	lower debris deposit	4535	50	3490–3470: 3375–3085: 3050–3035	−21.8	Bonsall 2000
Reindeer Cave, Inchnadamph	41	NC 2682 1704	OxA-5761	metatarsal	CNU/5, fissure fill	4720	50	3635–3485: 3475–3370	−20.9	Hedges et al.1998
			OxA-5759	juv. femur	CNU/3, fissure fill	4520	50	3370–3085: 3060–3030	−21.7	Hedges et al.1998
			OxA-5758	juv. humerus	CNU/2, fissure fill	4515	60	3490–3470: 3375–3020	−21.4	Hedges et al.1998
			OxA-5760	juvenile scapula	CNU/4, fissure fill	4470	50	3355–3010: 2980–2965: 2955–2940	−20.8	Hedges et al.1998
Robin Hood's Cave, Creswell	42	SK 5341 7419	OxA-7386	frontal	465, layer OB	5000	40	3945–3830: 3825–3695	−20.5	Hedges et al.1998
			OxA-1807	cervical vertebra	132, tip 68	4870	120	3950–3490: 3465–3375		Hedges et al. 1991
Scabba Wood Rock Shelter	43	SE 5269 0196	UB-3629	human bone	Individual A, orange brown loam	4590	30	3500–3435: 3380–3330: 3215–3185: 3160–3125		Chamberlain 1996

Site name	No.	NGR	Lab no.	Bone sampled	Ident. no. and context	Date BP	Error	Cal. date ranges BC (2 Σ)	δ13 C (‰)	Reference
Sewell's Cave	44	SD 7847 6658	OxA-13537	parietal	S.20, cave earth layer	5002	33	3945–3855: 3820–3700	−21.3	Griffiths 2011
Spurge Hole, Gower	45	SS 5468 8730	OxA-3815	femur	cave entrance	4830	100	3910–3880: 3805–3365	−19.8	Schulting and Richards 2002a
Thaw Head Cave	46	SD 7105 7590	OxA-14264	mandible	individual 1, cave floor	5040	31	3955–3760: 3725–3715	−21.0	Griffiths 2011
Tornewton Cave	47	SX 8172 6733	OxA-5864	lower right 2nd Incisor	individual 1	4680	60	3635–3550: 3540–3355		Hedges et al. 1997
Totty Pot, Somerset	48	ST 4825 5357	OxA-16458	adult left femur	TP6	4706	35	3635–3560: 3540–3485: 3475–3370	−21.0	Schulting, Gardiner et al. 2010
			OxA-16462	juvenile right ulna	TP 2004.9/419	4498	35	3355–3085	−21.1	Schulting, Gardiner et al. 2010
			OxA-16459	adult left femur	TP'63 2004.9/968	4473	35	3345–3080: 3070–3025	−21.2	Schulting, Gardiner et al. 2010
			OxA-16461	juvenile right femur	TP '63	4442	36	3335–3210: 3195–3150: 3140–3000: 2995–2925	−21.2	Schulting, Gardiner et al. 2010
			OxA-16460	juvenile left femur	TP 2004.9/257	4008	39	2835–2820: 2630–2460	−21.6	Schulting, Gardiner et al. 2010

Appendix 2: European caves with Neolithic human remains

Name	Source	Approx. start (BC)	Approx. end (BC)	Lat.	Long.
Belgium					
Abri des Autours I	Bocherens et al. 2007	4320	3980	50.27	4.91
Abri de Chauveau	Toussaint and Becker 1994	2400	1900	50.35	4.88
Abri de la Sigillee	Toussaint and Becker 1994	3380	2900	50.38	5.52
Abri du Pape	Bocherens et al. 2007	2900	2580	50.27	4.91
Abri Longue-Va	Toussaint and Becker 1994	2800	2300	50.50	5.18
Abri Masson	Toussaint and Becker 1994	3360	2450	50.49	5.61
Caverne AB, Hastiere	Bocherens et al. 2007	4040	3790	50.20	4.87
Caverne B, Hastiere	Bocherens et al. 2007	4220	3800	50.22	4.84
Caverne de Jausse	Bocherens et al. 2007	3020	2490	50.43	5.00
Caverne L, Hastiere	Bocherens et al. 2007	3980	3710	50.22	4.84
Caverne M, Hastiere	Bocherens et al. 2007	3320	2870	50.22	4.84
Caverne O, Hastiere	Bocherens et al. 2007	2890	2590	50.22	4.84
Caverne Q, Hastiere	Bocherens et al. 2007	3650	3100	50.22	4.84
Caverne Y, Hastiere	Bocherens et al. 2007	3310	2870	50.22	4.84
Chauveau CH1	Toussaint and Becker 1994	3900	3650	50.32	4.94
Fissure Jacques	Toussaint and Becker 1994	2980	2640	50.49	5.60
Grotte 1, Maurenne	Cauwe 2004	3625	3195	50.22	4.80
Grotte Ambre	Bocherens et al. 2007	2140	1680	50.14	4.73
Grotte Bibiche	Toussaint and Becker 1994	2880	2350	50.24	4.90
Grotte d'Anseremme	Cauwe 2004	3965	3640	50.24	4.90
Grotte de la Betche-aux-Rotches	Cauwe 2004	3010	2350	50.47	4.70
Grotte de la Cave	Toussaint and Becker 1994	2500	2000	50.22	4.80
Grotte des Dessins	Toussaint and Becker 1994	2800	2400	50.38	4.88
Grotte du Burnot	Toussaint and Becker 1994	2870	2300	50.36	4.87
Grotte du Coleoptere	Cauwe 2004	3635	3365	50.37	5.53
Grotte Humain	Polet et al. 2014	3080	2480	50.20	5.27

Name	Source	Approx. start (BC)	Approx. end (BC)	Lat.	Long.
Grotte Sepulcrale	Toussaint and Becker 1994	2800	2450	50.35	5.49
Grotte Triangulaire	Toussaint and Becker 1994	2820	2450	50.59	5.41
La Cave, Maurenne	Bocherens et al. 2007	3630	1950	50.22	4.82
Le Cemitiere, Hastiere	Bocherens et al. 2007	3030	2690	50.22	4.84
Ossuaire du Femur	Toussaint and Becker 1994	2770	2300	50.50	5.19
Petite Caverne, Hastiere	Bocherens et al. 2007	3090	2700	50.22	4.84
Roche aux Corneilles	Cauwe 2004	3340	2905	50.35	4.85
Sepulture 1 des Avins	Cauwe 2004	2490	2145	50.40	5.29
Sepulture 2 des Avins	Cauwe 2004	3350	2930	50.40	5.29
Sepulture 3 des Avins	Cauwe 2004	2885	2310	50.40	5.29
Trou de Han	Warmenbol 2014	3000	2500	50.12	5.18
Trou de la Heid	Toussaint and Becker 1994	3560	3360	50.48	5.58
Trou de la PJ	Bocherens et al. 2007	3050	2490	50.41	5.62
Trou des Blaireaux	Cauwe 2004	3000	2500	50.11	4.74
Trou des Nots	Toussaint and Becker 1994	3700	3500	50.31	4.82
Trou du Frontal	Toussaint and Becker 1994	3350	2950	50.21	4.95
Trou Fanfan	Bocherens et al. 2007	2880	2580	50.22	4.84
Trou Felix	Bocherens et al. 2007	3010	2670	50.27	4.91
Trou Garcon	Bocherens et al. 2007	2910	2620	50.22	4.84
Trou Reuviau	Cauwe 2004	3960	3670	50.20	4.83
Trou Rosette	Polet et al. 1996	2900	2600	50.21	4.95
Eastern Adriatic					
Ajdovska Jama	Bonsall et al. 2007	4340	4220	45.97	15.48
Badanj Cave	Forenbaher et al. 2010	4800	4000	43.10	17.90
Ciganska Jama	Trimmis 2016	4900	4815	45.66	14.89
Cina Turcului	Bonsall et al. 2015	6200	5800	44.59	22.26
Grapceva Cave	Forenbaher et al. 2013	4800	4300	43.16	16.65
Mala Triglavca	Mlekuž et al. 2008	4030	3800	45.85	15.67
Markova Cave	Forenbaher et al. 2010	4800	4000	43.18	16.43
Ravlica Pecina	Forenbaher et al. 2013	4800	4000	43.26	17.28
Vela Spilja (Korcula)	Forenbaher et al. 2013	4800	4000	42.93	16.89
Zelena Cave	Forenbaher et al. 2010	4800	4000	43.02	16.22
France					
Abri 2 de Fraischamp	Zemour 2011	5300	4900	43.98	5.10
Abri 3 de Chinchon	Zemour 2011	5300	4900	43.93	5.11
Abri 3 de Saint-Mitre	Zemour 2011	5300	4900	43.89	5.63

Appendix 2

Name	Source	Approx. start (BC)	Approx. end (BC)	Lat.	Long.
Abri de Bellefonds	Roussot–Larroque 1984	3100	2300	45.52	4.89
Abri de Cortiou	Zemour 2011	5300	4900	43.21	5.49
Abri de la Font-des-Pigeons	Zemour 2011	5300	4900	43.37	5.13
Abri de la Vessigne	Zemour 2011	5300	4900	44.35	4.58
Abri de Villeforceix	Roussot–Larroque 1984	3100	2300	45.98	0.99
Abri Edward	Zemour 2011	5300	4900	44.02	5.23
Aven de la Boucle	Vander Linden 2006	3635	3025	43.88	3.94
Baume Bourbon	Guilaine and Manen 2007	5300	4900	43.92	4.46
Baume Fontbregoua	Le Bras–Goude et al. 2010	5450	4800	43.55	6.23
Baumes–Chaudes	Vander Linden 2006	3400	2600	44.58	3.43
Can–Pey cave	Baills and Chaddaoui 1996	3500	2030	42.44	2.57
Cova de l'Esperit	Zemour 2011	5300	4900	42.85	2.85
Grotte d'Artenac	Roussot–Larroque 1984	3100	2300	45.84	0.34
Grotte d'Unang	Guilaine and Manen 2007	5300	4900	44.03	5.16
Grotte de Bois–Bertaud	Boulestin et al. 2002	3500	3000	45.66	−0.12
Grotte de Casse–Bartas	Beyneix 2012	3100	2300	44.41	1.03
Grotte de l'Adaouste	Le Bras–Goude et al. 2010	4800	4300	43.65	5.65
Grotte de la Gelie	Boulestin et al. 2002	3500	3000	45.49	0.35
Grotte de Leygonie	Roussot–Larroque 1984	3100	2300	45.09	0.49
Grotte de Riaux I	Zemour 2011	5300	4900	43.37	5.43
Grotte de Terraillic	Roussot–Larroque 1984	3100	2300	44.04	1.96
Grotte de Treilles 1	Herrscher et al. 2013	3100	2300	43.93	3.02
Grotte des Fadets	Boulestin et al. 2002	3500	3000	45.69	0.41
Grotte des Truels II	Laporte et al. 2011	3100	2300	44.10	3.10
Grotte du Cordonnier	Beyneix 2012	3100	2300	44.41	1.04
Grotte du Four	Roussot–Larroque 1984	3100	2300	44.23	1.72
Grotte du Gardon	Zemour 2011	5300	4900	45.98	5.40
Grotte du Mas de Jammes	Roussot–Larroque 1984	3100	2300	44.38	1.91
Grotte du Mazuc	Roussot–Larroque 1984	3100	2300	44.08	1.72
Grotte du Pas de Joulie	Laporte et al. 2011	3100	2300	44.08	3.39
Grotte du Queroy	Roussot–Larroque 1984	3100	2300	45.65	0.33
Grotte du Rudemont	Blaizot et al. 2001	4320	3000	49.02	6.02
Grotte du Trou Amiaut	Boulestin et al. 2002	3500	3000	45.80	0.30
Grotte du vieux-mounoi	Zemour 2011	5300	4900	43.30	5.88
Grotte Gazel	Guilaine and Manen 2007	5300	4900	43.31	2.41

Name	Source	Approx. start (BC)	Approx. end (BC)	Lat.	Long.
Grotte le Meunier	Beeching, forthcoming	2800	2300	44.31	4.56
Grotte Maison Blanche	Boulestin et al. 2002	3355	2935	45.74	0.34
Grotte Sicard	Zemour 2011	5300	4900	43.37	5.15
Grottes des Barbilloux	Beyneix 2012	4500	3700	45.19	0.49
L'Abri Araguina Sennola	Zemour 2011	5300	4900	45.56	−0.35
L'Abri du Moulin du Roc	Beyneix 2012	3100	2300	44.87	0.92
L'Abri du Pas–Estret	Beyneix 2012	4500	3700	44.91	1.12
L'Abri Jean Cros	Zemour 2011	5300	4900	43.06	2.47
L'Abri Pendimoun	Binder et al. 1993	5800	5600	43.81	7.51
La Baume des Maures	Vander Linden 2006	3400	2600	43.35	6.45
La Grotte Camprafaud 'Lacune'	Zemour 2011	5300	4900	43.48	2.90
La Grotte des Cramails	Beyneix 2012	3100	1350	44.77	0.26
La Grotte des Heritages	Zemour 2011	5300	4900	43.46	5.40
Laugerie Haute	Beyneix 2012	4500	3700	44.94	1.05
Le Rastel	Le Bras–Goude et al. 2010	4800	4300	43.77	7.38
Resplandy Cave	Vander Linden 2006	3400	2600	43.48	2.75
Roc de la Borie	Beyneix 2012	4500	3700	44.60	1.02
Trou de Vivies	Vander Linden 2006	3400	2600	43.17	2.96
Germany					
Blatterhohle	Orschiedt 2012	3800	3200	51.39	7.54
Felsstalle	Orschiedt 2012	3400	2500	48.42	9.86
Hohlenstein–Stadel	Orschiedt 2012	4470	4040	48.48	10.07
Jungfernhohle	Orschiedt 2012	5500	2500	49.99	10.98
Schellnacker Wand	Orschiedt 2012	3400	2500	48.94	11.79
Vogelherd	Orschiedt 2012	3890	2540	48.40	9.93
Great Britain					
An Corran Rock Shelter, Skye	this volume	3500	2300	57.64	−6.21
Ash Tree Cave, Derbys	this volume	2460	1915	53.28	−1.23
Ash Tree Shelter, Whitwell	this volume	3620	3370	53.28	−1.23
Backwell Cave, Somerset	this volume	3360	3090	51.41	−2.73
Blue John Cavern	this volume	2870	2580	53.35	−1.80
Bob's Cave, Devon	this volume	3970	3670	50.34	−4.01
Bower Farm, Staffordshire	this volume	3635	3375	52.77	−1.96
Broken Cavern, Devon	this volume	3940	3380	50.50	−3.67
Cae Gronw Cave	this volume	3620	2235	53.23	−3.48
Carding Mill Bay 1, Argyll	this volume	3690	2875	56.41	−5.49

Name	Source	Approx. start (BC)	Approx. end (BC)	Lat.	Long.
Cathole Cave	this volume	3630	3365	51.59	−4.11
Cattedown Cave, Devon	this volume	3935	3670	50.36	−4.12
Cave Ha 3, Yorks	this volume	3655	3040	54.09	−2.32
Chapel Cave	this volume	3695	3375	54.10	−2.18
Charterhouse Warren Farm Swallett	this volume	2455	1985	51.29	−2.73
Chelm's Combe, Somerset	this volume	3630	3365	51.29	−2.77
Darfar Crag Cave	this volume	3765	3365	53.10	−1.86
Flint Jack's Cave, Cheddar	this volume	3345	2915	51.28	−2.77
Foxhole Cave, Paviland	this volume	3790	3350	51.55	−4.25
George Rock Shelter, Goldsland	this volume	3950	3650	51.44	−3.28
Happaway Cave	this volume	3765	3635	50.47	−3.52
Hay Wood Cave, Somerset	this volume	3795	3385	51.32	−2.95
Jubilee Cave, Yorks	this volume	3695	3530	54.09	−2.25
Kent's Cavern, Devon	this volume	2035	1690	50.47	−3.50
King Arthur's Cave, Herefords	this volume	3635	3350	51.84	−2.66
Kinsey Cave, N. Yorks	this volume	3960	3790	54.09	−2.30
Lesser Kelco Cave, Yorks	this volume	3650	3520	54.08	−2.29
Little Hoyle Cave	this volume	3795	3495	51.67	−4.73
Little Quarry, Llandudno	this volume	3640	3360	53.33	−3.77
Markland Grips, Derbys	this volume	3710	3425	53.27	−1.24
Mother Grundy's Parlour, Creswell	this volume	3635	3110	53.26	−1.20
Nanna's Cave, Caldy Island	this volume	3490	3150	51.64	−4.68
North End Pot, N. Yorks	this volume	2885	2635	54.18	−2.49
Ogof–y–Benlog, Caldy Island	this volume	3630	3360	51.64	−4.68
Orchid Cave, Maeshafn	this volume	3010	2475	53.14	−3.20
Ossom's Crag Cave, Staffs	this volume	3910	3380	53.10	−1.86
Picken's Hole, Somerset	this volume	3696	3379	51.29	−2.87
Pontnewydd, Denbeighs.	this volume	3369	2930	53.23	−3.48
Raschoille Cave, Oban	this volume	3795	3200	56.40	−5.48
Reindeer Cave, Inchnadamph	this volume	3625	3025	58.11	−4.94
Robin Hood's Cave, Creswell	this volume	3955	3720	53.26	−1.20

Name	Source	Approx. start (BC)	Approx. end (BC)	Lat.	Long.
Scabba Wood Shelter, S. Yorks	this volume	3500	3125	53.51	−1.21
Sewell's Cave, Yorks	this volume	3940	3700	54.09	−2.33
Spurge Hole, Gower	this volume	3905	3370	51.57	−4.10
Thaw Head, Yorks	this volume	3950	3715	54.18	−2.45
Tornewton Cave, Devon	this volume	3635	3360	50.55	−3.67
Totty Pot, Somerset	this volume	3940	2940	51.28	−2.74
Greece					
Alepotrypa	Tomkins 2009	5500	3000	36.69	22.39
Ayia Triada	Demoule and Perlès 1993	4500	3000	38.10	24.34
Ayios Ioannis	Tomkins 2013	4500	3000	35.51	24.07
Eileithyia	Tomkins 2013	3300	2000	35.33	25.30
Ellenospilia	Tomkins 2013	4500	3900	35.59	23.77
Fournospilia	Demoule and Perlès 1993	4500	3000	38.87	22.30
Franchthi	Tomkins 2009	6500	3000	37.33	23.15
Gerani Cave	Tomkins 2013	6000	3900	35.40	24.61
Hagios Nikolaos	Demoule and Perlès 1993	4500	3000	38.57	21.02
Kalythies	Tomkins 2009	5500	3000	36.44	28.22
Kitsos	Tomkins 2009	5500	3000	37.74	24.04
Koumarospilio	Tomkins 2013	3300	3000	35.58	24.16
Limnon	Tomkins 2009	4900	4500	38.03	22.12
Marathon	Demoule and Perlès 1993	4500	3000	38.12	23.95
Prosymna Cave	Perlès 2001	6000	5300	37.67	22.73
Rodochori Cave	Demoule and Perlès 1993	4500	3000	40.56	22.28
Skaphidia	Tomkins 2013	3300	2000	35.34	25.13
Skotieni	Tomkins 2009	5500	3000	38.58	23.90
Stravomyti	Tomkins 2013	3600	3300	35.19	25.14
Tharounia	Demoule and Perlès 1993	4500	3000	38.46	23.93
Theopetra	Tomkins 2009	6500	5300	39.56	21.31
Trapeza	Tomkins 2013	3300	2000	35.19	25.51
Tsoungiza	Tomkins 2009	6500	5300	37.94	22.62
Iberian peninsula					
Algar do Barrao	Weiss–Krejci 2012	3640	3130	39.46	−8.65
Algar do Bom Santo	Weiss–Krejci 2012	3760	3115	39.22	−8.99
Bolores rock shelter	Weiss–Krejci 2012	2880	2355	39.21	−9.10
Camino del Molino	Weiss–Krejci 2012	3010	2355	38.06	−1.22

Appendix 2

Name	Source	Approx. start (BC)	Approx. end (BC)	Lat.	Long.
Cova das Lapas	Weiss–Krejci 2012	3500	3030	39.58	−8.94
Covao d'Almeidia	Weiss–Krejci 2012	3360	2940	40.11	−8.70
Covao do Poco	Weiss–Krejci 2012	3325	2885	39.52	−8.59
Cueva de Marizulo	Weiss–Krejci 2012	4315	3975	43.22	−1.98
Cueva de Nerja	Weiss–Krejci 2012	4825	4460	36.76	−3.85
Gruta da Feteira I	Weiss–Krejci 2012	3520	2495	39.26	−9.29
Gruta da Feteira II	Weiss–Krejci 2012	3695	2895	39.26	−9.29
Gruta do Cadaval	Weiss–Krejci 2012	4330	3800	39.65	−8.41
Gruta do Caldeirao	Weiss–Krejci 2012	5300	3635	39.65	−8.42
Gruta do Escoural	Weiss–Krejci 2012	3645	2910	38.54	−8.16
Gruta dos Alqueves	Weiss–Krejci 2012	3360	3025	40.20	−8.66
Gruta dos Ossos	Weiss–Krejci 2012	3635	2060	39.56	−8.54
Las Yurdinas II	Fernández–Crespo and de-la-Rúa 2016	3340	2780	42.62	−2.70
Llometes caves	Salazar–Garcia et al. 2016	4200	2800	38.70	−0.51
Nossa S. das Lapas	Weiss–Krejci 2012	5220	3650	39.66	−8.52
Pena Larga	Fernández–Crespo and de-la-Rúa 2016	3485	2750	42.61	−2.52
Pico Ramos	Weiss–Krejci 2012	3910	2350	43.33	−3.12
San Juan ante Portam Latinam	Weiss–Krejci 2012	3495	2495	42.52	−2.50
Ireland					
Annagh Cave	Dowd 2015	3700	3365	52.68	−8.45
Ballynamintra Cave	Dowd 2015	3315	2300	52.12	−7.76
Bantick Cave	Dowd 2015	3485	3035	52.81	−9.00
Bats' Cave	Dowd 2015	3335	2920	52.81	−9.00
Carrigmurrish Cave	Dowd 2015	3345	2945	52.11	−7.75
Connaberry Cave C	Dowd 2015	3640	3370	52.17	−8.46
Elderbush Cave	Dowd 2015	3690	2300	52.81	−9.00
Kilgreany cave	Dowd 2015	3795	2910	52.10	−7.74
Killavullen Cave 3	Dowd 2015	3370	3100	52.15	−8.52
Killura Cave	Dowd 2015	3630	3365	52.17	−8.54
Killuragh Cave	Fibiger 2016	3765	3410	52.60	−8.32
Knocknarea Cave C	Dowd 2015	3640	3375	54.26	−8.58
Knocknarea Cave K	Dowd 2015	3630	3035	54.26	−8.58
Oonaglour Cave	Dowd 2015	3360	3090	52.11	−7.77
Quinlan's Quarry Cave	Dowd 2015	3780	3640	52.11	−7.72
Red Cellar Cave	Dowd 2015	3625	3365	51.63	−8.52
Italy and Mediterranean Islands					
Arene Candide	Sparacello et al. 2016	5620	5470	44.16	8.33
Arma dell'Aquila	Sparacello et al. 2016	4980	4360	44.20	8.33

Name	Source	Approx. start (BC)	Approx. end (BC)	Lat.	Long.
Bergeggi	Sparacello et al. 2016	5500	5000	44.24	8.44
Boragni	Sparacello et al. 2016	5500	5000	44.22	8.36
Bur Mghez	Stoddart and Malone 2012	4100	3600	35.91	14.44
Cala Colombo	Robb 1994	3800	3400	41.09	17.00
Ghar Dalam	Zammitt 1930	4100	3600	35.84	14.53
Grotta Continenza	Robb 2007	5660	4240	41.96	13.54
Grotta dei Piccioni	Robb 1994	3600	3300	42.22	13.96
Grotta del Guano	Skeates 2012	3950	3550	40.27	9.42
Grotta del Leone	Robb 1994	3800	3400	44.01	10.27
Grotta dell'Orso	Robb 1994	5200	4900	42.99	11.85
Grotta delle Felci	Robb 1994	4500	3600	40.55	14.23
Grotta delle Mura	Robb 1994	5800	4800	40.95	17.31
Grotta delle Settecannelle	Robb 1994	5800	4800	42.54	11.76
Grotta di Porto Badisco	Robb 2007	4500	3600	40.08	18.48
Grotta di S. Angelo	Robb 1994	4500	3600	40.56	17.22
Grotta di Sa 'Ucca de Su Tintirriolu	Skeates 2012	3950	3550	40.45	8.65
Grotta di San Michele ai Cappuccini	Skeates 2012	4000	3200	40.58	9.00
Grotta di Sant'Angelo sulla Montagna dei Fiori	Robb 1994	5800	4800	42.75	13.62
Grotta La Cava	Robb 1994	4500	3600	42.02	13.52
Grotta Maritza	Robb 1994	4800	4400	42.01	13.54
Grotta Pacelli	Robb 1994	5800	3600	40.87	17.15
Grotta Patrizi	Robb 2007	5000	4500	42.06	12.40
Grotta Pavolella	Robb 2007	5800	5200	39.79	16.32
Grotta Refugio	Skeates 2012	4700	4000	40.27	9.43
Grotta Sa Rocca Ulari	Skeates 2012	4000	3200	40.52	8.74
Grotta Scaloria	Robb et al. 2015	5500	5000	41.64	15.91
Grotta Sisaia	Skeates 2012	2450	2050	40.25	9.47
Grotta Verde	Skeates 2012	5300	4700	40.56	8.16
Grutta I de Longu Fresu	Skeates 2012	4250	4050	39.85	9.27
Pian del Ciliegio	Sparacello et al. 2016	5500	5000	44.20	8.38
Pollera	Sparacello et al. 2016	5500	5000	44.20	8.31
Riparo sotto roccia Su Carroppu	Skeates 2012	5700	5300	39.21	8.56

References

Alberti, B. and Bray, T. 2009. Animating archaeology: of subjects, objects and alternative ontologies. *Cambridge Archaeological Journal* 19/3, 337–343.
Aldhouse-Green, S. and Peterson, R. 2007. The Goldsland Caves research project: excavations in 2007. *Archaeology in Wales* 47, 68–71.
Aldhouse-Green, S., Pettitt, P. and Stringer, C. 1996. Holocene humans at Pontnewydd and Cae Gronw Caves. *Antiquity* 70, 444–447.
Alt, K., Zesh, S., Garrido-Pena, R., Knipper, C., Szécsényi-Nagy, A., Roth, C., Tejedor-Rodriguez, C., Held, P., Garcia-Martinez-de-Lagrán, I., Navitainuck, D., Magallón, H. and Rojo-Guerra, M. 2016. A community in life and death: the Late Neolithic tomb at Alto de Reinoso (Burgos, Spain). *PLoS ONE* 11/1, 1–32.
Ambers, J. and Bowman, S. 2003. Radiocarbon measurements from the British Museum: datelist XXVI. *Archaeometry* 45/3, 531–540.
ApSimon, A. 1986. Picken's Hole, Compton Bishop, Somerset: early Devensian bear and wolf den and Middle Devensian hyaena den and Palaeolithic site. In Collcutt, S. (ed.) *The Palaeolithic of Britain and its Nearest Neighbours*. Sheffield: University of Sheffield Press, 55–56.
Armit, I. and Finlayson, B. 1992. Hunter-gatherers transformed: the transition to agriculture in Northern and Western Europe. *Antiquity* 66, 664–676.
Armstrong, A. 1956. Report on the excavation of Ash Tree Cave, near Whitwell, Derbyshire, 1949 to 1957. *Derbyshire Archaeological Journal* 76, 57–64.
Ashbee, P. 1966. The Fussell's Lodge long barrow excavations 1957. *Archaeologia* 100, 1–80.
Auden, W. H. 1962. *The Dyer's Hand and Other Essays*. London: Faber and Faber.
Bailey, D. 2001. *Balkan prehistory: Exclusion, Incorporation and Identity*. London and New York: Routledge.
Baills, H. and Chaddaoui, L. 1996. La sépulture collective de Can-Pey (Pyrénées-Orientales): Étude des pratiques funéraires. *Bulletins et Mémoires de la Société d'Anthropologie de Paris* NS 8/3–4, 365–371.
Balch, H. and Palmer, P. 1926. Excavations at Chelm's Combe, Cheddar. *Proceedings of the Somerset Natural History and Archaeological Society* 72/2, 93–123.
Barnatt, J. and Edmonds, M. 2002. Places apart? Caves and monuments in Neolithic and Earlier Bronze Age Britain. *Cambridge Archaeological Journal* 12/1, 113–129.

Barrett, J. 1988. Fields of discourse: reconstituting a social archaeology. *Critique of Anthropology* 7/3, 5–16.

Barrett, J. 2001. Agency, the duality of structure, and the problem of the archaeological record. In Hodder, I. (ed.) *Archaeological Theory Today*. Cambridge: Cambridge University Press.

Bateman, T. 1861. *Ten Years' Diggings in Celtic and Saxon Grave Hills*. London: J.R. Smith.

Bayliss, A. Benson, D., Galer, D., Humphrey, L., McFadyen, L. and Whittle, A. 2007. One thing after another: the date of the Ascott-under-Wychwood long barrow. *Cambridge Archaeological Journal* 17/1, 29–44.

Bayliss, A., Bronk Ramsey, C., van der Plicht, J. and Whittle, A. 2007. Bradshaw and Bayes: towards a timetable for the Neolithic. *Cambridge Archaeological Journal* 17/1, 1–28.

Bayliss, A., Whittle, A. and Wysocki, M. 2007. Talking about my generation: the date of the West Kennet long barrow. *Cambridge Archaeological Journal* 17/1, 85–101.

Beeching, A. 2002. La fin du Chasséen et le Néolithique final dans le bassin du Rhône moyen. In Ferrari, A. and Visentini, P. (eds) *Il Declino del Mondo Neolitico. Ricerche in Italia Centro-settentrionale fra aspetti Peninsulari, Occidentali e Nord-alpini*. Pordenone: Quaderni del Museo Archeologico del Friuli Occidentale and Museo delle Scienze e commune di Pordenone, 67–83.

Beeching, A. forthcoming. The Neolithic of Southern France. In Marcigny, C., McFadyen, L. and Roberts, J. (eds) *The Prehistory of France*. Cambridge: Cambridge University Press.

Beeching, A., Berger, J-F., Brochier, J.L., Ferber, F., Helmer, D. and Sidi Maamar, H. 2000. Chasséens: agriculteurs ou éleveurs, sédentaires ou nomads? Quels types de milieu, d'économies et de sociétés? In Leduc, M., Valdeyron, N. and Vaquer, J. (eds) *Sociétés et espaces. Actualité de la recherche. Rencontres Méridionales de Préhistoire Récente*. Toulouse: Archives d'Ecologie Préhistorique, 59–79.

Bennett, W. 1938. Giant's Graves, Pen-y-Ghent. *Yorkshire Archaeological Journal* 34, 318.

Beyneix, A. 2012. Le Monde des Morts au Néolithique en Aquitaine: essai de synthèse. *L'Anthropologie* 116, 222–233.

Binder, D., Bouchier, J-L., Duday, H., Helmer, D., Marinval, P., Thjiebault, S. and Wattez, J. 1993. L'Abri Pendimoun à Castellar (Alpe-Maritimes). Nouvelles données sur le complexe culturel de la céramique imprimée méditerranéenne dans son contexte stratigraphique. *Gallia Préhistoire* 35, 177–251.

Binder, D. and Maggi, R. 2001. Le Néolithique ancien de l'arc liguro-provençal. *Bulletin de la Société Préhistorique Française* 98/3, 411–422.

Bjerck, H.B. 2012. On the outer fringe of the human world: phenomenological perspectives on anthropomorphic cave paintings in Norway. In Bergsvik, K.A. and Skeates, R. (eds) *Caves in Context: The Cultural Significance of Caves and Rockshelters in Europe*. Oxford: Oxbow, 48–64.

Blaizot, F., Boës, X., Lalaï D., Le Meur, N. and Maigrot, Y. 2001. Premières données sur le traitement des corps humains à la transition du Néolithique récent et du Néolithique final dans le Bas-Rhin: dimensions culturelles. *Gallia Préhistoire* 43, 175–235.

Bloch, M. 1977. The past and the present in the present. *Man* (NS) 12/2, 278–292.

Bloch, M. 1982. Death, women and power. In Bloch, M. and Parry, J. (eds) *Death and the Regeneration of Life*. Cambridge: Cambridge University Press, 211–230.

Bloch, M. and Parry, J. 1982. Introduction: death and the regeneration of life. In Bloch, M. and Parry, J. (eds) *Death and the Regeneration of Life*. Cambridge: Cambridge University Press, 1–44.

Blockley, S. 2005. Two hiatuses in human bone radiocarbon dates in Britain (17000 to 5000 cal BP). *Antiquity* 79, 505–513.

Blockley, S. 2006. Living and dying in transition: funerary behaviour, subsistence and landscape use in Britain 16,000–6,000 Cal BP. Unpublished PhD thesis, University of Bradford.

Bocherens, H., Polet, C. and Toussaint, M. 2007. Palaeodiet of Mesolithic and Neolithic populations of Meuse Basin (Belgium): evidence from stable isotopes. *Journal of Archaeological Science* 34, 10–27.

Bogaard, A. and Halstead, P. 2015. Subsistence practices and social routine in Neolithic southern Europe. In Fowler, C., Harding J. and Hofmann, D. (eds) *The Oxford Handbook of Neolithic Europe*. Oxford: Oxford University Press, 385–410.

Bonsall, C. 2000. Oban – Raschoille (NM 8547 2888). *Discovery and Excavation in Scotland* 1999, 112.

Bonsall, C., Horvat, M., McSweeney, K., Masson, M., Higham, T., Pickard, C. and Cook, G. 2007. Chronological and dietary aspects of the human Burials from Ajdovska Cave, Slovenia. *Radiocarbon* 49/2, 727–740.

Bonsall, C., Pickard, C. and Ritchie, G.A. 2012. From Assynt to Oban: some observations on prehistoric cave use in western Scotland. In Bergsvik K.A. and Skeates R. (eds) *Caves in Context. The Cultural Significance of Caves and Rockshelters in Europe*. Oxford: Oxbow, 10–21.

Bonsall, C., Vasić, R., Boroneant, A., Roksandic, M., Soficaru, A., McSweeney, K., Evatt, A., Aguraiuja, Ű., Pickard, C., Dimitrijević, V., Higham, T., Hamilton, D. and Cook, G. 2015. New AMS 14C dates for human remains from Stone Age sites in the Iron Gates reach of the Danube, southeast Europe. *Radiocarbon* 57/1, 33–46.

Booth, T., Chamberlain, A. and Parker Pearson, M. 2015. Mummification in Bronze Age Britain. *Antiquity* 89/347, 1155–1173.

Boulestin, B., Gomez de Soto, J. and Laporte, L. 2002. La grotte sépulcrale du Néolithique récent de la Maison Blanche à Saint Project (Charente); premières observations. *Bulletin de la Société Préhistorique Française* 99/1, 39–47.

Bourdieu, P. 1977. *Outline of a Theory of Practice*. Cambridge: Cambridge University Press.

Bourdieu, P. 1984. *Distinction: A Social Critique of the Judgement of Taste*. Cambridge, MA: Harvard University Press.

Bourdieu, P. 1990. *The Logic of Practice*. Cambridge: Polity Press.

Brace, S., Diekmann, Y., Booth, T.J., Faltyskova, Z., Rohland, N., Mallick, S., Ferry, M., Michel, M., Oppenheimer, J., Broomandkhoshbacht, N., Stewardson, K., Walsh, S., Kayser, M., Schulting, R., Craig, O., Sheridan, A., Parker Pearson, M., Stringer, C., Reich, D., Thomas, M.G. and Barnes, I. in prep. Population replacement in Early Neolithic Britain. *bioRxiv: The Preprint Server for Biology*. Available at – https://www.biorxiv.org/content/early/2018/02/18/267443 (accessed 08/05/2018).

Bradley, R. 2000. *An Archaeology of Natural Places*. London and New York: Routledge.

Bréhard, S., Beeching, A. and Vigne, J-D. 2010. Shepherds, cowherds and site function on middle Neolithic sites of the Rhône Valley: An archaeozoological approach to the organization of territories and societies. *Journal of Anthropological Archaeology* 29, 179–188.

Bristow, J., Simms, Z. and Randolph-Quinney, P. 2011. Taphonomy. In Black, S. and Ferguson, E. (eds) *Forensic Anthropology 2000–2010*. Boca Raton: CRC Press, 279–317.

Bronk Ramsey, C. 2009. Bayesian analysis of radiocarbon dates. *Radiocarbon*, 51/1, 337–360.

Bronk Ramsey, C., Higham, T., Bowles, A. and Hedges, R. 2004. Improvements to the pretreatment of bone at Oxford. *Radiocarbon* 46/1, 155–163.

Bronk Ramsey, C., Higham, T., Brock, F., Baker, D. and Ditchfield, P. 2009. Radiocarbon dates from the Oxford AMS system: Archaeometry datelist 33. *Archaeometry* 51/2, 323–349.

Bronk Ramsey, C., Higham, T., Brock, F., Baker, D., Ditchfield, P. and Staff, R. 2015. Radiocarbon dates from the Oxford AMS system: Archaeometry datelist 35. *Archaeometry* 57/1, 177–216.

Bronk Ramsey, C., Pettitt, P., Hedges, R., Hodgins, G. and Owen, D. 2000. Radiocarbon dates from the Oxford AMS system: Archaeometry datelist 30. *Archaeometry* 42/2, 459–479.

Brown, F. and Clark, P. 2011. Stainton West (Parcel 27) CNDR, Cumbria: Post-excavation assessment. Oxford Archaeology (North): unpublished client report.

Buckland, P., Chamberlain, A., Collins, P., Dungworth, D., Frederick, C., Merrony, C., Nystrom, P. and Parker Pearson, M. 1998. *Scabba Wood: Interim Report on Excavations, April 1998*. University of Sheffield: unpublished archive report.

Buikstra, J. and Beck, L. 2006. *Bioarchaeology: The Contextual Analysis of Human Remains*. Boston: Academic Press.

Callander, J.G., Cree, J.E. and Ritchie, J. 1927. Preliminary report on caves containing Palaeolithic relics near Inchnadamph, Sutherland. *Proceedings of the Society of Antiquaries of Scotland* 61, 169–172.

Campbell, J.B. 1977. *The Upper Palaeolithic of Britain. A Study of Man and Nature in the Late Ice Age*. Oxford: Clarendon Press.

Cane, C. and Cane J. 1986. The excavation of a Mesolithic cave site near Rudgeley, Staffordshire. *Staffordshire Archaeological Studies* 3, 1–12.

Canilao, M. 2012. Three burial coffin traditions in upland Llocos Sur. *The Cordillera Review* 4/1, 47–68.

Cassen, S. 1993. Material culture and chronology of the Middle Neolithic of western France. *Oxford Journal of Archaeology* 12/2, 197–208.

Cauwe, N. 2004. Les sépultures collectives néolithiques en grotte du bassin Mosan. Bilan documentaire. *Anthropologica et Praehistorica* 115, 217–224.

Celino, S. 1990. Death and burial practices and beliefs of the Cordilleras. Unpublished EdD dissertation, University of Baguio.

Chadwick, A.M. 1992. *An Excavation of a Rock Shelter Burial in Scabba Wood, Sprotbrough, South Yorkshire*. South Yorkshire Archaeology Service: unpublished client report.

Chamberlain, A. 1996. More dating evidence for human remains in British caves. *Antiquity* 70, 950–953.

Chamberlain, A. 2014. *Gazetteer of Caves, Fissures and Rock Shelters in Britain Containing Human Remains.* Revised Version 2014. Online resource hosted by the University of Bristol Spelaeological Society. Available at – http://caveburial.ubss.org.uk/ (accessed 16/09/2016).

Chamberlain, A. and Ray, K. 1994. *A Catalogue of Quaternary Fossil-Bearing Cave Sites in the Plymouth Area.* Plymouth Archaeology occasional publication 1. Plymouth: Plymouth City Council.

Chapman, J. 1999. Deliberate house-burning in the prehistory of central and eastern Europe. *Glyfer och Arkeologiska Rum-en Vänbok* 3, 113–122.

Charles, R. and Jacobi, R. 1994. The lateglacial fauna from the Robin Hood Cave, Creswell Crags: a re-assessment. *Oxford Journal of Archaeology* 13, 1–32.

Collcutt, S. 1984. The sediments. In Green, H.S. (ed.) *Pontnewydd Cave: a Lower Palaeolithic Hominid site in Wales: The First Report.* Cardiff: National Museum of Wales, 31–76.

Compton, C.H. 1892. Proceedings of the association. *Journal of the British Archaeological Association* 48/1, 75–92.

Conard, N., Grootes, P. and Smith, F. 2004. Unexpectedly recent dates for human remains from Vogelherd. *Nature* 430, 198–201.

Conneller, C. 2010. *An Archaeology of Materials: Substantial Transformations in Early Prehistoric Europe.* London: Routledge.

Connock, K. 1985. Rescue excavation of the ossuary remains at Raschoille Cave, Oban. Lorn Archaeological and Historical Society: unpublished excavation report.

Connock, K., Finlayson, W. and Mills, C. 1993. Excavation of a shell midden site at Carding Mill Bay, near Oban, Scotland. *Glasgow Archaeological Journal* 17, 25–38.

Cooper, R., Ryder, P. and Solman, K. 1976. The North Yorkshire windypits: a review. *Transactions of the British Cave Research Association* 3/2, 77–94.

Crombé, P. and Robinson, E. 2014. ^{14}C dates as demographic proxies in Neolithicisation models of northwestern Europe: a critical assessment using Belgium and northeast France as a case study. *Journal of Archaeological Science* 52, 558–566.

Cummings, V. 2017. *The Neolithic of Britain and Ireland.* London and New York: Routledge.

Cummings, V. and Harris, O. 2011. Animals, people and places: The continuity of hunting and gathering practices across the Mesolithic-Neolithic transition in Britain. *European Journal of Archaeology* 14/3, 361–382.

Cummings, V., Midgley, M. and Scarre, C. 2015. Chambered tombs and passage graves of western and northern Europe. In Fowler, C., Harding J. and Hofmann, D. (eds) *The Oxford Handbook of Neolithic Europe.* Oxford: Oxford University Press, 813–838.

Dabkowski, J. 2014. High potential of calcareous tufas for integrative multidisciplinary studies and prospects for archaeology in Europe. *Journal of Archaeological Science* 52, 72–83.

D'Anna, A. 1995. Le Néolithique final en Provence. In Voruz, J-L. (ed.) *Chronologies néolithiques, de 6000 à 2000 avant notre ère dans le bassin rhodanien*. Actes du Colloque d'Ambérieu-en-Bugey. Documents du Départment d'Anthropologie Université de Genève, 20, 265–286.

Davies, M. 1981. Identification of bones from Orchid Cave, Maeshafn, Clwyd. Unpublished assessment report archived on Ebbs, C (ed.) *Caves of North Wales*. Available at – https://docs.google.com/viewer?a=v&pid=sites&srcid =ZGVmYXVsdGRvbWFpbnxjYXZlc29mbm9ydGh3YWxlc3xneDo1O GI0MTg1MTc2MDZjMTQ3 (accessed 24/07/2017).

Davies, M. 1989a. Recent advances in cave archaeology in south west Wales. In Ford, T. (ed.) *Limestone and Caves of Wales*. Cambridge: Cambridge University Press, 79–91.

Davies, M. 1989b. Cave archaeology in North Wales. In Ford, T. (ed.) *Limestone and Caves of Wales*. Cambridge: Cambridge University Press, 92–101.

Davies, P. and Lewis, J. 2004. A Late Mesolithic/Early Neolithic site at Langley's Lane, near Midsomer Norton, Somerset. *Past* 49, 7–8.

Davis, O. and Sharples, N. 2017. Early Neolithic enclosures in Wales: a review of the evidence in light of recent discoveries at Caerau, Cardiff. *The Antiquaries Journal* 97, 1–26.

Dearne, M. and Lord, T.C. 1998. *The Romano-British Archaeology of Victoria Cave, Settle: Researches into the Site and its Artefacts*. BAR British Series 273. Oxford: British Archaeological Reports.

De Landa, M. 2006. *A New Philosophy of Society: Assemblage Theory and Social Complexity*. London and New York: Continuum.

Delhon, C., Thiébault, S. and Berger, J-F. 2009. Environment and landscape management during the Middle Neolithic in Southern France: evidence for agro-sylvo-pastoral systems in the Middle Rhône Valley. *Quaternary International* 200, 50–65.

Demoule, J-P. and Perlès, C. 1993. The Greek Neolithic: a new review. *Journal of World Prehistory* 7/4, 355–416.

Dinnis, R., Bello, S.M., Chamberlain, A., Coleman, C. and Stringer, C. 2014. A cut-marked Neolithic human tooth from Ash Tree Shelter, Derbyshire. *Cave and Karst Science* 41/3, 114–117.

Dirks, P., Berger, L., Roberts, E., Kramers, J., Hawks, J., Randolph-Quinney, P., Elliott, M., Musiba, C., Churchill, S., de Ruiter, D., Schmid, P., Backwell, L., Belyanin, G., Boshoff, P., Hunter, L., Feuerriegel, E., Gurtov, A., Harrison, J., Hunter, R., Kruger, A., Morris, H., Makhubela, T., Peixotto, B. and Tucker, S. 2015. Geological and taphonomic context for the new hominin species *Homo naledi* from the Dinaledi Chamber, South Africa. *eLife* 2015/4, e09561. DOI:http://dx.doi.org/10.7554/eLife.09561

Dobres, M-A. 1995. Gender and prehistoric technology: on the social agency of technical strategies. *World Archaeology* 27/1, 25–49.

Dowd, M. 2008. The use of caves for funerary and ritual practices in Neolithic Ireland. *Antiquity* 82, 305–317.

Dowd, M. 2015. *The Archaeology of Caves in Ireland*. Oxford: Oxbow.

Duday, H. 2006. Archaeothanatology or the archaeology of death. In Gowland, R. and Knüsel, C. (eds) *Social Archaeology of Funerary Remains*. Oxford: Oxbow, 30–56.

Dunbar, L. and Thoms, J. 2008. Human remains in cave deposits at Benderloch, Oban. Summary report. AOC Archaeology Group: unpublished client report.

Durkheim, E. 1995. *The elementary forms of religious life* (trans. K. Fields). New York: The Free Press.

Ebbs, C. 2017. *Caves of North Wales: an information resource.* Available at – https ://sites.google.com/site/cavesofnorthwales/07-caves-o (accessed 24/07/2017).

Edmonds, M. 1999. *Ancestral Geographies of the Neolithic: Landscapes, Monuments and Memory.* London and New York: Routledge.

Evans, E. and Lewis, R. 2003. *The prehistoric funerary and ritual monument survey of Glamorgan and Gwent: overviews.* Swansea: Glamorgan-Gwent Archaeological Trust.

Evans-Pritchard, E. 1940. *The Nuer: A Description of the Modes of Livelihood and Political Institutions of a Nilotic People.* Oxford: Clarendon Press.

Everton, A. and Everton, R. 1972. Hay Wood Cave burials, Mendip Hills, Somerset. *Proceedings of the University of Bristol Spelaeological Society* 13/1, 5–29.

Fairchild, I. and Baker, A. 2012. *Speleothem Science: From Process to Past Environments.* Oxford: Wiley-Blackwell.

Fernández-Crespo, T. and de-la-Rúa, C. 2016. Demographic differences between funerary caves and megalithic graves of Northern Spanish Late Neolithic/Early Chalcolithic. *American Journal of Physical Anthropology* 160, 284–297.

Fibiger, L. 2016. Osteoarchaeological analysis of human skeletal remains from 23 Irish caves. In Dowd, M. (ed.) *Underground Archaeology: Studies on Human Bones and Artefacts from Ireland's Caves.* Oxford: Oxbow.

Fitzpatrick, A. 2011. *The Amesbury Archer and the Boscombe Bowmen.* Salisbury: Wessex Archaeology.

Ford, T. 2001. *Sediments in caves.* BCRA Caves Studies Series 9. Buxton: British Cave Research Association.

Forenbaher, S., Kaiser, T. and Frame, S. 2010. Adriatic Mortuary Ritual at Grapčeva Cave, Croatia. *Journal of Field Archaeology* 35/4, 337–354.

Forenbaher, S., Kaiser, T. and Miracle, P. 2013. Dating the East Adriatic Neolithic. *European Journal of Archaeology* 16/4, 589–609.

Forenbaher, S. and Miracle, P. 2013. Transition to farming in the Adriatic: a view from the Eastern Shore. In Guilaine, J., Manen C. and Perrin T. (eds) *The Neolithic Transition in the Mediterranean.* Paris: Editions Errance.

Foucault, M. 1979. *Discipline and Punish: The Birth of the Prison* (trans. A. Sheridan). Harmondsworth: Peregrine Books.

Fowler C. 2003. Rates of (ex)change: Decay and growth, memory and the transformation of the dead in Early Neolithic Southern Britain. In Williams, H. (ed.) *Archaeologies of Remembrance: Death and MemoryiIn Past Societies.* New York: Kluwer Academic/Plenum, 45–63.

Fowler, C. 2013. *The Emergent Past: A Relational Realist Archaeology of Early Bronze Age Mortuary Practices.* Oxford: Oxford University Press.

Fowler, C., Harding, J. and Hofmann, D. (eds) 2015. *The Oxford Handbook of Neolithic Europe.* Oxford: Oxford University Press.

Fowler, C. and Harris, O. 2015. Enduring relations: exploring a paradox of new materialism. *Journal of Material Culture* 20/2, 127–148.

Gardiner, P. 2016. Totty Pot, Cheddar Somerset: a history of the archaeological excavations and finds from 1960 to 1998. *Proceedings of the University of Bristol Spelaeological Society* 27/1, 39–72.

Gardner, A. 2004. Introduction: social agency, power and being human. In Gardner, A. (ed.) *Agency Uncovered: Archaeological Perspectives on Social Agency, Power, and Being Human*. London: UCL Press, 1–15.
Gell, A. 1992. *The Anthropology of Time*. Oxford: Berg.
Gell, A. 1996. Vogel's net: traps as artworks and artworks as traps. *Journal of Material Culture* 1/1, 15–38.
Gell, A. 1998. *Art and Agency: An Anthropological Theory*. Oxford: Clarendon Press.
Gibson, A. 1994. Excavations at the Sarn-y-bryn-caled cursus complex, Welshpool, Powys, and the timber circles of Great Britain and Ireland. *Proceedings of the Prehistoric Society* 60, 143–223.
Gibson, A. and Bayliss, A. 2010. Recent work on the Neolithic round barrows of the upper Great Wold Valley, Yorkshire. In Leary, J., Darvill, T. and Field, D. (eds) *Round Mounds and Monumentality in the British Neolithic and Beyond*. Oxford: Oxbow, 72–107.
Giddens, A. 1979. *Central Problems in Social Theory: Action, Structure and Contradiction in Social Analysis*. London: Macmillan.
Giddens, A. 1984. *The Constitution of Society: Outline of the Theory of Structuration*. Cambridge: Polity Press.
Gilks, J. 1995. Later Neolithic and Bronze Age pottery from Thaw Head Cave, Ingleton, North Yorkshire. *Transactions of the Hunter Archaeological Society* 18, 1–11.
Gilks, J. and Lord, T. 1985. A Late Neolithic crevice burial from Selside, Ribblesdale, North Yorkshire. *Yorkshire Archaeological Journal* 57, 1–5.
Gilks, J. and Lord, T. 1993. A Neolithic antler macehead from North End Pot, Ingleton, North Yorkshire. *Transactions of the Hunter Archaeological Society* 17, 57–59.
Gillings, M. and Pollard, J. 1999. Non-portable stone artefacts and contexts of meaning: the tale of Grey Wether (www.musuems.ncl.ac.uk/Avebury/stone4.htm). *World Archaeology* 31/2, 179–193.
González-Fortes, G., Jones, E.R., Lightfoot, E., Bonsall, C., Lazar, C., Grandald'Anglade, A., Garralda, M.D., Drak, L., Siska, V., Simalcsik, A., Boroneanţ, A., Romaní, J.R.V., Rodríguez, M.V., Arias, P., Pinhasi, R., Manica, A. and Hofreiter, M. 2017. Paleogenomic evidence for multi-generational mixing between Neolithic farmers and Mesolithic hunter-gatherers in the Lower Danube basin. *Current Biology* 27/12, 1801–1810.e10. Available at – https ://doi.org/10.1016/j.cub.2017.05.023 (accessed 20/04/2018).
Gosselain, O. 1999. In pots we trust: the processing of clay and symbols in sub-Saharan Africa. *Journal of Material Culture* 4/2, 205–230.
Gowlett, J., Hall, E., Hedges, R. and Perry, C. 1986. Radiocarbon dates from the Oxford AMS system: archaeometry datelist 3. *Archaeometry* 28/1, 116–125.
Green, H.S. 1986. Excavations at Little Hoyle (Longbury Bank), Wales, in 1984. In Roe, D. (ed.) *Studies in the Upper Palaeolithic of Britain and Northwest Europe*. Oxford: British Archaeological Reports International Series 296, 99–120.
Gregory, R., Roberts, J., Robinson, M. and Shimwell, D. 2000. A retrospective assessment of 19th-century finds from a Little Orme quarry. *Archaeology in Wales* 40, 3–8.

Griffiths, S. 2011. Chronological modelling of the Mesolithic-Neolithic transition in Britain. Unpublished PhD thesis, Cardiff University.

Griffiths, S. 2014a. A Bayesian radiocarbon chronology of the early Neolithic of Yorkshire and Humberside. *The Archaeological Journal* 171/1, 2–29.

Griffiths, S. 2014b. Points in time: the chronology of rod microliths. *Oxford Journal of Archaeology* 33/3, 221–43.

Griffiths, S. 2016. What the chronology of early cereal domesticates tells us about the nature of the early Neolithic in Britain. Paper presented at the Neolithic Studies Group.

Gronenborn, D. 1999. A variation on a basic theme: the transition to farming in southern Central Europe. *Journal of World Prehistory* 13/2, 123–210.

Gronenborn, D. 2007. Beyond the models: 'Neolithisation' in Central Europe. In Whittle, A. and Cummings, V. (eds) *Going Over: The Mesolithic–Neolithic Transition in North-Western Europe (Proceedings of the British Academy 144)*. Oxford: Oxford University Press, 73–98.

Gronenborn, D. and Dolukhanov, P. 2015. Early Neolithic manifestations in Central and Eastern Europe. In Fowler, C., Harding J. and Hofmann, D. (eds) *The Oxford Handbook of Neolithic Europe*. Oxford: Oxford University Press, 195–214.

Guilaine, J. 2015. The Neolithization of Mediterranean Europe. In Fowler, C., Harding J. and Hofmann, D. (eds) *The Oxford Handbook of Neolithic Europe*. Oxford: Oxford University Press, 81–98.

Guilaine, J. and Manen, C. 2007. From Mesolithic to Early Neolithic in the western Mediterranean. In Whittle, A. and Cummings, V. (eds) *Going Over: The Mesolithic–Neolithic Transition in North-Western Europe (Proceedings of the British Academy 144)*. Oxford: Oxford University Press, 21–51.

Guilbert, G. 1982. Orchid Cave. *Archaeology in Wales* 22, 15.

Gutherz, X. and Jallot, L. 1995. Le Néolithique final du Languedoc méditerranéen. In Voruz, J-L. (ed.) *Chronologies néolithiques, de 6000 à 2000 avant notre ère dans le bassin rhodanien*. Actes du Colloque d'Ambérieu-en-Bugey. Documents du Départment d'Anthropologie Université de Genève, 20, 265–286.

Haak, W., Balanovsky, O., Sanchez, J.J., Koshel, S., Zaporozhchenko, V., Adler, C.J., Der Sarkissian, C.S.I., Brandt, G., Schwarz, C., Nicklisch, N., Dreseley, V., Fritsch, B., Balanovska, E., Villems, R., Meller, H., Alt, K.W., Cooper, A. and the Genographic Consortium. 2010. Ancient DNA from European Early Neolithic Farmers reveals their Near Eastern Affinities. *PLOS Biology* [online] 8. DOI:10.1371/journal. pbio.1000536. Available at – http://journals.plos.org/plosbiology/article?id=10.1371/journal.pbio.1000536 (accessed 20/04/2018).

Hachem, L. 2000. New observations on the Bandkeramik house and social organization. *Antiquity* 74, 308–312.

Haglund, W. 1992. Contribution of rodents to post-mortem artifacts of bone and soft tissue. *Journal of Forensic Sciences* 37/6, 1459–1465.

Haglund, W. 1997. Dogs and coyotes: post-mortem involvement with human remains. In Haglund, W. and Sorg, M. (eds) *Forensic Taphonomy: The Post-Mortem Fate of Human Remains*. Boca Raton: CRC Press, 367–381.

Haglund, W., Reay, D. and Swindler, D. 1988. Tooth mark artefacts and survival of bones in animal scavenged human skeletons. *Journal of Forensic Sciences* 33/4, 985–997.

Haglund, W. and Sorg, M. (eds) 1997. *Forensic Taphonomy: The Post-Mortem Fate of Human Remains*. Boca Raton: CRC Press.

Haglund, W. and Sorg, M. (eds) 2002. *Advances in Forensic Taphonomy: Method, Theory and Archaeological Perspectives*. Boca Raton: CRC Press.

Harman, G. 2009. *Prince of Networks: Bruno Latour and Metaphysics*. Melbourne: Re.Press.

Harris, O. 2017. Assemblages and scale in archaeology. *Cambridge Archaeological Journal* 27/1, 127–139.

Harris, O. and Cipolla, C. 2017. *Archaeological Theory in the New Millennium: Introducing Current Perspectives*. Oxford: Routledge.

Harris, O. and Robb, J. 2012. Multiple ontologies and the problem of the body in history. *American Anthropologist* 114/4, 668–679.

Hawkes, C., Rogers, J. and Tratman, E. 1978. Romano-British cemetery in the fourth chamber of Wookey Hole Cave, Somerset. *Proceedings of the University of Bristol Spelaeological Society* 15/1, 23–52.

Hedges, R., Housley, R., Bronk, C. and Van Klinken, G.J. 1991. Radiocarbon dates from the Oxford AMS system: archaeometry datelist 13. *Archaeometry* 33/2, 279–296.

Hedges, R., Housley, R., Bronk Ramsey, C. and Van Klinken, G.J. 1993. Radiocarbon dates from the Oxford AMS system: archaeometry datelist 17. *Archaeometry* 35/2, 305–326.

Hedges, R.E.M., Housely, R., Law, I. and Bronk, C.R. 1989. Radiocarbon dates from the Oxford AMS system: archaeometry datelist 9. *Archaeometry* 31/2, 207–234.

Hedges, R., Pettitt, P., Bronk Ramsey, C. and van Klinken, G. 1996. Radiocarbon dates from the Oxford AMS system: archaeometry datelist 22. *Archaeometry* 38/2, 391–419.

Hedges, R., Pettitt, P., Bronk Ramsey, C. and van Klinken, G. 1997. Radiocarbon dates from the Oxford AMS system: archaeometry datelist 24. *Archaeometry* 39/2, 445–471.

Hedges, R., Pettitt, P., Bronk Ramsey, C. and van Klinken, G. 1998. Radiocarbon dates from the Oxford AMS system: archaeometry datelist 26. *Archaeometry* 40/2, 437–455.

Hedges, R., Saville, A. and O'Connell, T. 2008. Characterizing the diet of individuals at the Neolithic chambered tomb of Hazleton North, Gloucestershire, England, using stable isotope analysis. *Archaeometry* 50/1, 114–128.

Heidegger, M. 1971. *Poetry, Language, Thought* (trans A. Hofstadter). New York: Harper and Row.

Hellewell, E. and Milner, N. 2011. Burial practices at the Mesolithic-Neolithic transition in Britain: change or continuity? *Documenta Praehistorica* 37, 61–68.

Herrscher, E., L'heureux, J., Goude, G., Dabernat, H. and Duranthon, F. 2013. Les pratiques de subsistance de la population Néolithique final de la grotte I des Treilles (commune de Saint-Jean-et-Saint-Paul, Aveyron). *Préhistoires Méditerranéennes* [on-line], 4. Available at – http://pm.revues.org/783 (accessed 02/10/2016).

Hertz, R. 1960. A contribution to the study of the collective representation of death. From *Death and the Right Hand* (trans. Rodney and Claudia Needham). Reprinted in Robben, A. (ed.) *Death, Mourning and Burial: A Cross-Cultural Reader.* Oxford: Blackwell, 197–212.

Higham, T., Bronk Ramsey, C., Brock, F., Baker, D. and Ditchfield, P. 2007. Radiocarbon dates from the Oxford AMS system: Archaeometry datelist 32. *Archaeometry* 49/S2, S1–S60.

Higham, T., Bronk Ramsey, C., Brock, F., Baker, D. and Ditchfield, P. 2011. Radiocarbon dates from the Oxford AMS system: Archaeometry datelist 34. *Archaeometry* 53/5, 1067–1084.

Hodder, I. 2012. *Entangled: An Archaeology of the Relationships Between Humans and Things.* Oxford: Wiley-Blackwell.

Hofmann, D. 2015. What have genetics ever done for us? The implications of aDNA data for interpreting identity in Early Neolithic Central Europe. *European Journal of Archaeology* 18/3, 454–476.

Holderness, H., Davies, G., Chamberlain, A. and Donahue, R. 2006. *Research Report – A Conservation Audit of Archaeological Cave Resources in the Peak District* and *Yorkshire Dales.* English Heritage Research Report 743.b. Sheffield: Archaeological Research and Consultancy at the University of Sheffield.

Ingold, T. 1993. The temporality of the landscape. *World Archaeology* 25/2, 152–174.

Ingold, T. 2000. *The Perception of the Environment: Essays in Livelihood, Dwelling and Skill.* London and New York: Routledge.

Ingold, T. 2007. Materials against materiality. *Archaeological Dialogues* 14/1, 1–16.

Ingold, T. 2011. *Being Alive: Essays In Movement, Knowledge and Description.* London: Routledge.

Jackson. J.W. 1953. Archaeology and palaeontology. In Cullingford, C. (ed.) *British Caving – An Introduction to Speleology.* London: Routledge and Kegan Paul, 252–346.

Jackson, J.W. and Mattinson, W. 1932. A cave on Giggleswick Scars, near Settle, Yorkshire. *The Naturalist,* 5–9.

Jennings, J. 1985. *Karst Geomorphology.* Oxford: Blackwell.

Jones, E.R., Zarina, G., Moiseyev, V., Lightfoot, E., Nigst, P.R., Manica, A., Pinhasi, R. and Bradley, D.G. 2017. The Neolithic transition in the Baltic was not driven by admixture with Early European Farmers. *Current Biology* 27/4, 576–582. Available at – https://doi.org/10.1016/j.cub.2016.12.060 (accessed 20/04/2018).

Joy, J. 2009. Reinvigorating object biography: reproducing the drama of object lives. *World Archaeology* 41/4, 540–556.

Keiller, A., Piggott, S., Passmore, A. and Cave, A. 1938. Excavation of an untouched chamber in the Lanhill long barrow. *Proceedings of the Prehistoric Society* 5, 122–150.

Kind, C-J. 1987. *Das Felsstalle: Eine jungpaläolithisch-frühmesolithische Abri-Station bei Ehingen-Muhlen, Alb-Donau-Kreis. Die Grabungen 1975–1980.* Stuttgart: Konrad Theiss Verlag.

King, A. 1970. *Early Pennine Settlement: A Field Study.* Clapham: Dalesman Publishing.

Kitson, E. 1931. A study of the Negro skull with special reference to the crania from Kenya colony. *Biometrika* 23, 271–314.

Knüsel, C. 2010. Bioarchaeology: a synthetic approach. *Bulletins et Mémoires de la Société d'Anthropologie de Paris* 22/1–2, 62–73.

Knüsel, C. 2014. Crouching in fear: terms of engagement for funerary remains. *Journal of Social Archaeology* 14/1, 26–58.

Knüsel, C. and Outram, A. 2006. Fragmentation of the body: comestibles, compost, or customary rite? In Gowland R. and Knüsel C. (eds) *Social Archaeology of Funerary Remains*. Oxford: Oxbow, 253–278.

Kopytoff, I. 1986. The cultural biography of things: commoditization as process. In Appadurai, A. (ed.) *The Social Life of Things*. Cambridge: Cambridge University Press, 64–91.

Kreuz, A., Märkle, T., Marinova, E., Rösch, M., Schäfer, E., Schamuhn, S. and Zerl, T. 2014. The Late Neolithic Michelsberg culture – just ramparts and ditches? A supraregional comparison of agricultural and environmental data. *Praehistoriche Zeitschrift* 89/1, 72–115.

Kusimba, C. and Kusimba, S. 2000. Hinterlands and cities: archaeological investigations of economy and trade in Tsavo, south-eastern Kenya. *Nyame Akuma* 54, 13–24.

Kusimba, C., Kusimba, S. and Wright, D. 2005. The development and collapse of precolonial ethnic mosaics in Tsavo, Kenya. *Journal of African Archaeology* 3/2, 243–265.

Lacaille, A.D. and Grimes, W.F. 1956. The prehistory of Caldey. *Archaeologia Cambrensis* 104, 85–165.

Lambeck, K. 1995. Late Devensian and Holocene shorelines of the British Isles and North Sea from models of glacio-hydro-isostatic rebound. *Journal of the Geological Society of London* 152, 437–448.

Laporte, L., Jallot, L. and Sohn, M. 2011. Mégalithismes en France. *Gallia Préhistoire* 53, 289–338.

Latour, B. 1999. *Pandora's hope: essays in the reality of science studies*. Cambridge, MA: Harvard University Press.

Latour, B. 2005. *Reassembling the Social: An Introduction to Actor-Network-Theory*. Oxford: Oxford University Press.

Lawson, T. 1981. The 1926–7 excavations of the Creag nan Uamh bone caves, near Inchnadamph, Sutherland. *Proceedings of the Society of Antiquaries of Scotland* 111, 7–20.

Le Bras-Goude, G., Binder, D., Zemour, A. and Richards, M. 2010. New radiocarbon dates and isotope analysis of Neolithic human and animal bone from the Fontbrégoua Cave (Salernes, Var, France). *Journal of Anthropological Sciences* 88, 167–178.

Leach, E. 1961. *Rethinking Anthropology*. London: Athlone Press.

Leach, S. 2006. Going underground: taphonomic and anthropological reanalysis of human skeletal remains from caves in northern Yorkshire. Unpublished PhD thesis, University of Winchester.

Leach, S. 2008. Odd one out? Earlier Neolithic deposition of human remains in caves and rock shelters in the Yorkshire Dales. In Murphy, E. (ed.) *Deviant Burial in the Archaeological Record*. Oxford: Oxbow, 35–56.

Lemonnier, P. 1993. The eel and the Ankave-Anga of Papua New-Guinea: material and symbolic aspects of trapping. In Hladik, C-M., Hladik, A., Linares, O., Pagezy, H., Semple, A. and Hadley, M. (eds) *Tropical Forests, People and Foods: Biocultural Interactions and Applications to Development*. London: Taylor and Francis, 673–682.

Lemonnier, P. 2012. *Mundane Objects: Materiality and Non-Verbal Communication*. Walnut Creek, CA: Left Coast Press.

Leroi-Gourhan, A. 1994. *Gesture and Speech* (trans. A. Bostock Berger). Cambridge, MA: The MIT Press. (Original publication 1964. *La geste et le parole*. Paris: Editions Albin Michel).

Levi-Strauss, C. 1963. *Structural Anthropology*. New York: Basic Books.

Levitan, B., Audsley, A., Hawkes, C., Moody, A., Moody, P., Smart, P. and Thomas, J. 1988. Charterhouse Warren Farm Swallet, Mendip, Somerset: exploration, geomorphology, taphonomy and archaeology. *Proceedings of the University of Bristol Spelaeological Society* 18/2, 171–239.

Levitan, B. and Smart, P. 1989. Charterhouse Warren Farm Swallet, Mendip, Somerset. Radiocarbon dating evidence. *Proceedings of the University of Bristol Spelaeological Society* 18/3, 390–394.

Lord, T., O'Connor, T., Siebrandt, D. and Jacobi, R. 2007. People and large carnivores as biostratinomic agents in Lateglacial cave assemblages. *Journal of Quaternary Science* 22/7, 681–694.

Louwe Kooijmans, L. 2007. The gradual transition to farming in the lower Rhine basin. In Whittle, A. and Cummings, V. (eds) *Going Over: The Mesolithic–Neolithic Transition in North-Western Europe (Proceedings of the British Academy 144)*. Oxford: Oxford University Press, 287–309.

Lyman, L. and Fox, G. 1997. A critical evaluation of bone weathering as an indication of bone assemblage formation. In Haglund, W. and Sorg, M. (eds) *Forensic Taphonomy: The Post-Mortem Fate of Human Remains*. Boca Raton: CRC Press, 223–248.

Madgwick, R., Redknap, M. and Davies, B. 2016. Illuminating Lesser Garth Cave, Cardiff: the human remains and post-Roman archaeology in context. *Archaeologia Cambrensis* 165, 201–229.

Marshall, Y. and Gosden, C. 1999. A cultural biography of objects. *World Archaeology* 31/2, 169–178.

McFadyen, L. 2006. Making architecture. In Benson, D. and Whittle, A. (eds) *Building Memories: The Neolithic Cotswold Long Barrow at Ascott-under-Wychwood, Oxfordshire*. Oxford: Oxbow, 348–354.

Meadows, J., Barclay, A. and Bayliss, A. 2007. A short passage of time: the dating of the Hazleton long cairn revisited. *Cambridge Archaeological Journal* 17/1, 45–64.

Meiklejohn, C., Chamberlain, A. and Schulting, R. 2011. Radiocarbon dating of Mesolithic human remains in Great Britain. *Mesolithic Miscellany* 21/2, 20–58.

Meiklejohn, C., Merrett, D., Nolan, R., Richards, M. and Mellars, P. 2005. Spatial relationships, dating and taphonomy of the human bone from the Mesolithic site of Conc Coig, Oronsay, Argyll, Scotland. *Proceedings of the Prehistoric Society* 71, 85–106.

Mellars, P. 1987. *Excavations on Oronsay: Prehistoric Human Ecology on a Small Island*. Edinburgh: Edinburgh University Press.

Mellars, P. and Wilkinson, M. 1980. Fish otoliths as indicators of seasonality in prehistoric shell middens: the evidence from Oronsay (Inner Hebrides). *Proceedings of the Prehistoric Society* 46, 19–44.

Mellor, D. 1981. *Real Time*. Cambridge: Cambridge University Press.

Merleau-Ponty, M. 1962. *The Phenomenology of Perception* (trans. C. Smith). London: Routledge and Kegan Paul.

Metcalf, P. and Huntington, R. 1991. *Celebrations of Death: The Anthropology of Mortuary Ritual* (2nd edition), Cambridge: Cambridge University Press.

Milner, N. and Craig, O. 2009. Mysteries of the middens: change and continuity across the Mesolithic–Neolithic transition. In Allen, M., Sharples, N. and O'Connor, T. (eds) *Land and People: Papers in Memory of John G. Evans*. Oxford: Oxbow (Prehistoric Society Research Paper No 2), 169–180.

Mlekuž, D. 2005. The ethnography of the Cyclops: Neolithic pastoralists in the eastern Adriatic. *Documenta Praehistorica* 32, 15–51.

Mlekuž, D. 2011. What can bodies do? Bodies and caves in the Karst Neolithic. *Documenta Praehistorica* 38, 97–108.

Mlekuž, D. 2012. Notes from the underground: caves and people in the Mesolithic and Neolithic karst. In. Bergsvik, K.A. and Skeates, R. (eds) *Caves in Context: The Cultural Significance of Caves and Rockshelters in Europe*. Oxford: Oxbow, 199–211.

Mlekuž, D., Budja, M., Payton, R., Bonsall, C. and Gašparič, A. 2008. Reassessing the Mesolithic/Neolithic 'gap' in Southeast European cave sequences. *Documenta Praehistorica* 35, 237–251.

Morris, J. 2011. *Investigating Animal Burials: Ritual, Mundane and Beyond*. BAR British Series 535. Oxford: Archaeopress.

Mörseburg, A., Alt, K. and Knipper, C. 2015. Same old in Middle Neolithic diets? A stable isotope study of bone collagen from the burial community of Jechtingen, southwest Germany. *Journal of Anthropological Archaeology* 39, 210–221.

Mourne, R., Case, D., Viles, H. and Bull, P. 2012. The sedimentary sequence. In Aldhouse-Green, S., Peterson, R. and Walker, E. (eds) *Neanderthals in Wales: Pontnewydd and the Elwy Valley caves*. Oxford: Oxbow, 48–67.

National Trust HBSMR 2003. *Spurge Hole Cave – Pennard West Cliff, Pennard and Bishopston*. Available at – http://archaeologydataservice.ac.uk/archsearch/record.jsf?titleId=1776005 (accessed 07/07/2017).

Needham, S. 2005. Transforming Beaker culture in north-west Europe: processes of fusion and fission. *Proceedings of the Prehistoric Society* 71, 171–217.

Nixon, D. 2011. Human and faunal remains from Blue John Cavern, Castleton, Derbyshire, UK. *Cave and Karst Science* 38/2, 93–95.

Northern Archaeological Associates 2002. Archaeological watching brief: Embsay water main, Langcliffe. Unpublished client report.

Oakley, K. 1958. The antiquity of the skulls reputed to be from Flint Jack's Cave, Cheddar, Somerset. *Proceedings of the University of Bristol Spelaeological Society* 8/2, 77–82.

Oosterbeek, L. 1997. Back home! Neolithic life and the rituals of death in the Portuguese Ribatejo. In Bonsall, C. and Tolan-Smith, C. (eds) *The human*

Use of Caves. British Archaeological Reports International Series 667. Oxford: Archaeopress, 70–78.

Orschiedt, J. 2012. Cave burials in prehistoric Central Europe. In Bergsvik, K.A. and Skeates, R. (eds) *Caves in Context: The Cultural Significance of Caves and Rockshelters in Europe.* Oxford: Oxbow, 212–224.

Panchal, J. and Cimacio, M. 2016. Culture Shock – a study of domestic tourists in Sagada, Philippines. *4th Interdisciplinary Tourism Research Conference*, Bodrum, Turkey, 334–338.

Parker Pearson, M. 2003. *The Archaeology of Death and Burial.* Stroud: Sutton.

Parker Pearson, M., Chamberlain, A., Craig, O., Marshall, P., Mulville, J., Smith, H., Chernery, C., Collins, M., Cook, G., Craig, G., Evans, J., Hiller, J., Montgomery, J., Schwenninger, J-L., Taylor, G. and Wess, T. 2005. Evidence for mummification in Bronze Age Britain. *Antiquity* 79/305, 529–546.

Patrick, L. 1985. Is there an archaeological record? In Schiffer, M. (ed.) *Advances in Archaeological Method and Theory 8.* New York: Academic Press.

Pels, P. 1998. The spirit of matter. On fetish, rarity, fact, and fancy. In Spyer, P. (ed.) *Border Fetishisms. Material Objects in Unstable Spaces.* London: Routledge, 91–121.

Pentecost, A., Thorpe, P., Harkness, D. and Lord, T. 1990. Some radiocarbon dates for tufas of the Carven district of Yorkshire. *Radiocarbon* 32, 93–97.

Perlès, C. 2001. *The Early Neolithic in Greece: The First Farming Communities in Europe.* Cambridge: Cambridge University Press.

Perrin, T. 2003. Mesolithic and Neolithic cultures co-existing in the upper Rhône valley. *Antiquity* 77, 732–739.

Peterson, R. 2004. Away from the numbers: diversity and invisibility in late Neolithic Wales. In Cummings, V. and Fowler, C. (eds) *The Neolithic of the Irish Sea: Materiality and Traditions of Practice.* Oxford: Oxbow, 191–201.

Peterson, R. 2013. Social memory and ritual performance. *Journal of Social Archaeology* 13/2, 266–283.

Peterson, R. 2018. Do caves have agency? In Büster, L., Mlekuž, D. and Warmenbol, E. (eds) *Between Worlds: Understanding Ritual Cave Use in Later Prehistory.* New York: Springer.

Picpican, I. 2003. *The Igorot mummies: A Socio-Cultural and Historical Treatise.* Quezon City: Rex Bookstore.

Polet, C. and Cauwe, N. 2007. Étude anthropologique des sépultures préhistoriques de l'abri des Autours (Province de Namur, Belgique). *Anthropologica et Praehistorica* 118, 71–110.

Polet, C., Dutour, O., Orban, R., Ivan, J. and Louryan, S. 1996. A healed wound caused by a flint arrowhead in a Neolithic human innominate from the Trou Rosette (Furfooz, Belgium). *International Journal of Osteoarchaeology* 6, 414–420.

Polet, C., Warmenbol, E., Carels, E. and Déom, H. 2014. La grotte sépulcrale de Humain (Marche-en-Famenne, B). Les restes humains et le goblet campaniforme du Néolithique récent/final. *Notae Prehistoricae* 34, 115–124.

Pollard, A. 1990. Down through the ages: a review of the Oban cave deposits. *Scottish Archaeological Review* 7, 58–74.

Provost, S., Binder, D., Duday, H., Durrenmath, G., Goude, G., Gourichon, L., Delhon, C., Gentile, I., Vuillien, M. and Zemour, A. 2017. A collective grave from the 6th to 5th millennia transition BCE: Mougins – Les Bréguières

(Alpes-Maritimes, France). *Gallia Préhistoire* [Online], 57. Available at – http://journals.openedition.org/galliap/591 (accessed 20/04/2018).

Pryor, F. 1998. *Etton: Excavations at a Neolithic Causewayed Enclosure near Maxey Cambridgeshire, 1982–7*. English Heritage Archaeological Report 18. London: English Heritage.

Quinney, P. 2000. Paradigms lost: changing interpretations of hominid behaviour patterns since ODK. In Rowley-Conwy, P. (ed.) *Animal Bones, Human Societies*. Oxford: Oxbow, 12–19.

Raistrick, A. 1936. Excavations at Sewell's Cave, Settle, W. Yorkshire. *Proceedings of the University of Durham Philosophical Society* 9, 191–204.

Reilly, S. 2003. Processing the dead in Neolithic Orkney. *Oxford Journal of Archaeology* 22/2, 133–154.

Reimer, P.J., Bard, E., Bayliss, A., Beck, J.W., Blackwell, P.G., Bronk Ramsey, C., Grootes, P.M., Guilderson, T.P., Haflidason, H., Hajdas, I., Hattž, C., Heaton, T.J., Hoffmann, D.L., Hogg, A.G., Hughen, K.A., Kaiser, K.F., Kromer, B., Manning, S.W., Niu, M., Reimer, R.W., Richards, D.A., Scott, E.M., Southon, J.R., Staff, R.A., Turney, C.S.M. and van der Plicht, J. 2013. IntCal13 and Marine13 Radiocarbon Age Calibration Curves 0–50,000 Years cal BP. *Radiocarbon* 55/4, 1869–1887.

Renfrew, C. 1979. *Investigations in Orkney. Reports of the Research Committee of the Society of Antiquaries of London 38*. London: Thames and Hudson.

Reynolds, Ff. 2014. A site's history does not end: transforming place through community archaeology at Tinkinswood chambered tomb and surrounding landscape, Vale of Glamorgan. *Journal of Community Archaeology and Heritage* 1/2, 173–189.

Richards, M. and Mellars, P. 1998. Stable isotopes and the seasonality of the Oronsay middens. *Antiquity* 72, 178–184.

Richards, M. and Schulting, R. 2006. Against the grain? A response to Milner et al. (2004). *Antiquity* 80, 444–456.

Richards, M. and Sheridan A. 2000. New AMS dates on human bone from Mesolithic Oronsay. *Antiquity* 74, 313–315.

Ricoeur, P. 1988. *Time and Narrative, Volume 3*. Chicago: Chicago University Press.

Robb, J. 1994. Burial and social reproduction in the Peninsular Italian Neolithic. *Journal of Mediterranean Archaeology* 7/1, 27–71.

Robb, J. 2007. *The Early Mediterranean Village: Agency, Material Culture, and Social Change in Neolithic Italy*. Cambridge: Cambridge University Press.

Robb, J. 2013. Material culture, landscapes of action, and emergent causation: a new model for the origins of the European Neolithic. *Current Anthropology* 54/6, 657–673.

Robb, J., Elster, E., Isetti, E., Knüsel, C., Tafuri, M.A. and Traverso, A. 2015. Cleaning the dead: Neolithic ritual processing of human bone at Scaloria Cave, Italy. *Antiquity* 89/1, 39–54.

Roberts, A. 1996. Evidence for Late Pleistocene and Early Holocene human activity and environmental change from the Torbryan Valley, South Devon. In Charman, D.J., Newnham R.M. and Croot D.G. (eds) *The Quaternary of Devon and East Cornwall: Field Guide*. London: Quaternary Research Association, 168–204.

Robinson, D. 2017. Assemblage theory and the capacity to value: an archaeological approach from Cache Cave, California, USA. *Cambridge Archaeological Journal* 27/1, 155–168.
Rojo-Guerra, M. and Garrido-Pena, R. 2012. From pits to megaliths: Neolithic burials in the interior of Iberia. In Gibaja J., Carvalho A. and Chambon P. (eds) *Funerary Practices in the Iberian Peninsula from the Mesolithic to the Chalcolithic*. BAR International Series 2417. Oxford: Archaeopress, 21–28.
Roksandic, M. 2002. Position of skeletal remains as a key to understanding mortuary behaviour. In Haglund, W. and Sorg, M. (eds) *Advances in Forensic Taphonomy: Method, Theory and Archaeological Perspectives*. Boca Raton: CRC Press, 99–118.
Rorty, R. 1985. Texts and lumps. *New Literary History* 17, 1–15.
Rorty, R. 1991. *Essays on Heidegger and Others: Philosophical Papers Volume 2*. Cambridge: Cambridge University Press.
Rosen, C. 2016. The use of caves during the Mesolithic in south west Britain. Unpublished PhD thesis, University of Worcester.
Roussot-Larroque, J. 1984. Artenac aujourd'hui: pour une nouvelle approche de l'énéolithisation de la France. *Revue Archéologique du Centre de la France* 23/2, 135–196.
Salazar-García, D., García-Puchal, O., Paz de Miguel-Ibáñez, M. and Talamo, S. 2016. Earliest evidence of Neolithic collective burials from eastern Iberia: radiocarbon dating at the archaeological site of Les Llometes (Alicante, Spain). *Radiocarbon* 58/3, 679–692.
Saville, A. 1990. *Hazleton North, Gloucestershire, 1979–82: The Excavation of a Neolithic Long Cairn of the Cotswold-Severn Group*. London: English Heritage.
Saville, A. 1998. An Corran, Staffin, Skye. *Discovery and Excavation in Scotland*, 126–127.
Saville, A. 2005. Archaeology and the Creag nan Uamh bone caves, Assynt, Highland. *Proceedings of the Society of Antiquaries of Scotland* 135, 343–369.
Saville, A., Hardy, K., Miket, R. and Ballin, T B. 2012. An Corran, Staffin, Skye: a rockshelter with Mesolithic and later occupation. *Scottish Archaeological Internet Reports* [online] 51. Society of Antiquaries of Scotland: Edinburgh. Available at – https://doi.org/10.5284/1017938 (accessed 30/10/2018).
Scarre, C. 2002. Contexts of monumentalism: regional diversity at the Neolithic transition in North-West France. *Oxford Journal of Archaeology* 21/1, 23–61.
Schiffer, M. 1976. *Behavioural Archaeology*. New York: Academic Press.
Schiffer, M., Skibo, J., Griffitts, J., Hollenback, K. and Longacre, W. 2001. Behavioural archaeology and the study of technology. *American Antiquity* 66/4, 729–737.
Schofield, P. and Vannan, A. 2014. Ingleborough and Clapham Commons, Craven, North Yorkshire: archaeological survey report. Unpublished client report for Yorkshire Peat Partnership. Available at – https://library.thehumanjourney.net/2455/1/L10674_Ingleborough%20Report_Full.pdf (accessed 06/06/2018).
Schulting, R. 2007. Non-monumental burial in Britain: a (largely) cavernous view. In Larsson, L., Lüth, F. and Terberger, T. (eds) *Non-Megalithic Mortuary Practices in the Baltic – New Methods and Research into the Development of Stone Age Society*. Schwerin: Bericht der Römisch-Germanischen Kommission 88, 581–603.

Schulting, R., Chapman, M. and Chapman, E.J. 2013. AMS 14C dating and stable isotope (carbon, nitrogen) analysis of an earlier Neolithic human skeletal assemblage from Hay Wood Cave, Mendip, Somerset. *Proceedings of the University of Bristol Spelaeological Society* 26/1, 9–26.

Schulting, R., Gardiner, P., Hawkes, C. and Murray, E. 2010. The Mesolithic and Neolithic human bone assemblage from Totty Pot, Cheddar, Somerset. *Proceedings of the University of Bristol Spelaeological Society* 25/1, 75–95.

Schulting, R. and Gonzales, M. 2008. 'Prestatyn Woman' reconsidered. In Bell, M. (ed.) *Prehistoric Coastal Communities: The Mesolithic in Western Britain*. CBA Research Report 149. York: Council for British Archaeology, 303–305.

Schulting, R. and Richards, M. 2002a. Finding the coastal Mesolithic in southwest Britain: AMS dates and stable isotope results on human remains from Caldey Island, south Wales. *Antiquity* 76, 1011–1025.

Schulting, R. and Richards, M. 2002b. The wet, the wild and the domesticated: the Mesolithic-Neolithic transition on the west coast of Scotland. *European Journal of Archaeology* 5/2, 147–189.

Schulting, R., Sheridan, A., Crozier, R. and Murphy, E. 2010. Revisiting Quanterness: new AMS dates and stable isotope data from an Orcadian chamber tomb. *Proceedings of the Society of Antiquaries of Scotland* 140, 1–50.

Schulting, R. and Wysocki, M. 2005. 'In this chambered tumulus were found cleft skulls': an assessment of the evidence for cranial trauma in the British Neolithic. *Proceedings of the Prehistoric Society* 71, 107–138.

Scott, K. 1986. Man in Britain in the late Devensian: evidence from Ossom's Cave. In Roe, D. (ed.) *Studies in the Upper Palaeolithic of Britain and North West Europe*. Oxford: British Archaeological Reports, 63–87.

Seip, L. 1999. Transformations of meaning: the life history of a Nuxalk mask. *World Archaeology* 31/2, 272–287.

Shanks, M. 2007. Symmetrical archaeology. *World Archaeology* 39/4, 589–596.

Shanks, M. and Tilley, C. 1987. *Social Theory in Archaeology*. Oxford: Polity Press.

Simmons, T. 2002. Taphonomy of a karstic cave execution site at Hrgar, Bosnia-Herzegovina. In Haglund, W. and Sorg, M. (eds) *Advances In Forensic Taphonomy: Method, Theory and Archaeological Perspectives*. Boca Raton: CRC Press, 263–275.

Simmons, T., Cross, P., Adlam, R. and Moffatt, C. 2010. The influence of insects on decomposition rate in buried and surface remains. *Journal of Forensic Sciences* 55/4, 889–892.

Simpson, E. 1950. The Kelcow Caves, Giggleswick, Yorkshire. *Cave Science: Journal of the British Speleological Association* 2, 258–262.

Skeates, R. 2005. *Visual Culture and Archaeology: Art and Social Life in Prehistoric South-East Italy*. Duckworth: London.

Skeates, R. 2012. Caves in need of context: Prehistoric Sardinia. In Bergsvik, K.A. and Skeates, R. (eds) *Caves in Context: The Cultural Significance of Caves and Rockshelters in Europe*. Oxford: Oxbow, 166–187.

Skeates, R. 2013. Constructed Caves: transformations of the underworld in prehistoric southeast Italy. In Moyes, H. (ed.) *Sacred Darkness: A Global Perspective on the Ritual Use of Caves*. Boulder: University Press of Colorado, 27–44.

Smith, I. 2012a. Kirkhead Cavern, Kent's Bank Cavern and Whitton's Cave near Allithwaite – geology, sediments and archaeology. In O'Regan, H., Faulkner, T. and Smith, I. (eds) *Cave Archaeology and Karst Geomorphology of North West England: Field Guide*. London: Quaternary Research Association/British Cave Research Association, 98–102.

Smith, I. 2012b. Radio-carbon dating of bones from caves in Furness. *Cumberland and Westmorland Antiquarian and Archaeological Society Newsletter* 70, 6.

Sparacello, V.S., Roberts, C., Canci, A., Moggi-Cecchi, J. and Marchi, D. 2016. Insights on the paleoepidemiology of ancient tuberculosis from the structural analysis of postcranial remains from the Ligurian Neolithic (northwesten Italy). *International Journal of Paleopathology* 15, 50–64.

Stevens, C. and Fuller, D. 2012. Did Neolithic farming fail? The case for a Bronze Age agricultural revolution in the British Isles. *Antiquity* 86, 707–722.

Stoddart, S. and Malone, C. 2013. Caves of the living, caves of the dead: experiences above and below ground in Prehistoric Malta. In Moyes, H. (ed.) *Sacred Darkness: A Global Perspective on the Ritual Use of Caves*. Boulder: University Press of Colorado, 45–58.

Taylor, T., Lord, T. and O'Connor, T. 2011. Recent work at Kinsey Cave. University of Bradford: Unpublished archive report.

Tellier, G. 2009. What information can be gained from the analysis of human teeth on the nature, lifestyle and mortuary practices of the Neolithic mortuary population from Goldsland, Vale of Glamorgan? Unpublished BSc dissertation, University of Central Lancashire.

Thomas, D. and Britnell, W. 2008. Nant Hall Road. In Bell, M. (ed.) *Prehistoric coastal communities: the Mesolithic in Western Britain*. CBA Research Report 149. York: Council for British Archaeology, 267–269.

Thomas, J. 1996. *Time, Culture and Identity: An Interpretive Archaeology*. London: Routledge.

Thomas, J. 2003. Thoughts on the 'repacked' Neolithic revolution. *Antiquity* 77, 67–74.

Thomas, J. 2013. *The Birth of Neolithic Britain: An Interpretive Account*. Oxford: Oxford University Press.

Thomas, J. and Whittle, A. 1986. Anatomy of a Tomb: West Kennet revisited. *Oxford Journal of Archaeology* 5/2, 129–156.

Thurnam, J. 1860. On the examination of a chambered long barrow at West Kennet, Wiltshire. *Archaeologia* 38, 405–421.

Tomkins, P. 2009. Domesticity by default. Ritual, ritualization and cave use in the Neolithic Aegean. *Oxford Journal of Archaeology* 28/2, 125–153.

Tomkins, P. 2013. Landscapes of ritual, identity, and memory: reconsidering Neolithic and Bronze Age cave use in Crete, Greece. In Moyes, H. (ed.) *Sacred Darkness: A Global Perspective on the Ritual Use of Caves*. Boulder: University Press of Colorado, 59–80.

Toussaint, M. and Becker, A. 1994. Une sépulture du Michelsberg: Le trou de la Heid à Comblain-au-Pont (Province de Liège, Belgique). *Bulletin de la Société Préhistorique Française* 91/1, 77–84.

Tratman, E.K. 1938. The excavation of Backwell Cave, Somerset. *Proceedings of the University of Bristol Speleological Society* 5/1, 57–74.

Tratman, E.K. 1964. Picken's Hole, Crook Peak, Somerset. A Pleistocene site. Preliminary note. *Proceedings of the University of Bristol Spelaeological Society* 10/2, 112–115.

Trimmis, K. 2016. *Atlas of Balkan Neolithic Caves.* Available at – http://balkan-cavearchaeology.weebly.com/ (accessed 11/04/2017).

Vander Linden, M. 2006. For whom the bell tolls: Social hierarchy vs social integration in the Bell Beaker culture of southern France (third millennium BC). *Cambridge Archaeological Journal* 16/3, 317–332.

Vanmontfort, B. 2008. Forager–farmer connections in an 'unoccupied' land: First contact on the western edge of LBK territory. *Journal of Anthropological Archaeology* 27/2, 149–160.

Van Nèdervelde, J. and Davies, M. 1976. Nanna's Cave. *Archaeology in Wales* 16, 24.

Villa, P., Courtin, J., Helmer, D., Shipman, P., Bouville, C., Mahieu, E., Belluomini, G. and Branca, M. 1986. Un cas de cannibalisme au Néolithique: Boucherie et rejet de restes humains et animaux dans la grotte de Fontbrégoua à Salernes (Var). *Gallia Préhistoire* 29/1, 143–171.

Walker, E., Case, D., Ingrem, C., Jones, J. and Mourne R. 2014. Excavations at Cathole Cave, Gower, Swansea. *Proceedings of the University of Bristol Spelaeological Society* 26/2, 131–169.

Walsh, S., Knüsel, C. and Melton, N. 2011. A re-appraisal of the Early Neolithic human remains excavated at Sumburgh, Shetland, in 1977. *Proceedings of the Society of Antiquaries of Scotland* 141, 3–17.

Waltham, T. and Murphy, P. 2013. Cave geomorphology. In Waltham, T. and Lowe, D. (eds) *Caves and Karst of the Yorkshire Dales, Volume 1.* Buxton: British Cave Research Association, 117–146.

Ward, J. 1916. The St Nicholas chambered tumulus, Glamorgan II. *Archaeologia Cambrensis* 16, 239–294.

Warmenbol, E. 2014. Le 'Trou de Han' à Han-sur-Lesse. In Frébutte, C. (ed.) *Coup d'Oeil sur 25 Ans de Recherches Archéologiques à Rochefort, de 1989 à 2014.* Namur: Institut du Patrimoine Wallon, 68–81.

Weight, W. 2002. *Hydrogeology Field Manual* (2nd edition). New York: McGraw Hill.

Weiss-Krejci, E. 2012. Shedding light on dark places: Deposition of the dead in caves and cave-like features in Neolithic and Coper Age Iberia. In Bergsvik, K.A. and Skeates, R. (eds) *Caves in Context: The Cultural Significance of Caves and Rockshelters in Europe.* Oxford: Oxbow, 118–137.

Whitehouse, N., Schulting, R., McClatchie, M., Barratt, P., McLaughlin, R., Bogaard, A., Colledge, S., Marchant, R., Gaffrey, J. and Bunting, J. 2013. Neolithic agriculture on the European Western Frontier: the boom and bust of early farming in Ireland. *Journal of Archaeological Science* 51, 181–205.

Whitehouse, R. 2015. Water turned to stone: stalagmites and stalactites in cult caves in prehistoric Italy. *Accordia Research Papers* 14 (2014–2015), 49–62.

Whittle, A. 1996. *Europe in the Neolithic: The Making of New Worlds.* Cambridge: Cambridge University Press.

Whittle, A., Barclay, A., Bayliss, A., McFadyen, L., Schulting, R. and Wysocki, M. 2007. Building for the dead: events, processes and changing worldviews from the thirty-eighth to the thirty-fourth centuries cal. BC in southern Britain. *Cambridge Archaeological Journal* 17/1, 123–147.

Whittle, A., Bayliss, A. and Wysocki, M. 2007. Once in a lifetime: the date of the Wayland's Smithy long barrow. *Cambridge Archaeological Journal* 17/1, 103–121.

Whittle, A., Healy, F. and Bayliss, A. 2011. *Gathering time: dating the early Neolithic enclosures of Southern Britain and Ireland*. Oxford: Oxbow.

Whittle, A. and Wysocki, M. 1998. Parc le Breos Cwm transepted long cairn, Gower, West Glamorgan: date, contents, and context. *Proceedings of the Prehistoric Society* 64, 139–182.

Williams, D., Richardson, J. and Richardson, R. 1987. Mesolithic sites at Malham Tarn and Great Close Mire, North Yorkshire. *Proceedings of the Prehistoric Society* 53, 363–383.

Williams, G. 2008. What types of mortuary practice are represented in the human bone assemblage of Goldsland Cave? A report on site G: skeletal analysis, interpretation and discussion. Unpublished BSc dissertation, University of Central Lancashire.

Woodman, P. 2015. *Ireland's First Settlers: Time and the Mesolithic*. Oxford: Oxbow.

Worth, R.N. 1887. On the discovery of human remains in a Devonshire bone cave. *Transactions of the Royal Geological Society of Cornwall* 11, 105–112.

Wysocki, M., Bayliss, A. and Whittle, A. 2007. Serious mortality: the date of the Fussell's Lodge long barrow. *Cambridge Archaeological Journal* 17/1, 65–84.

Wysocki, M. and Whittle, A. 2000. Diversity, lifestyles and rites: new biological and archaeological evidence from British Earlier Neolithic mortuary assemblages. *Antiquity* 74, 591–601.

Wysocki, M., Griffiths, S., Hedges, R., Bayliss, A., Higham, T., Fernandez-Jalvo, Y. and Whittle, A. 2013. Dates, diet, and dismemberment: evidence from the Coldrum Megalithic Monument, Kent. *Proceedings of the Prehistoric Society* 79, 1–30.

Zammitt, T. 1930. The prehistoric remains of the Maltese Islands. *Antiquity* 4, 55–79.

Zemour, A. 2011. Les pratiques funéraires au début du Néolithique en Méditerranée nord-occidentale sont-elles homogènes? In Sénépart, I., Perrin, T., Thirault, É. and Bonnardin, S. (eds), *Actes des 8èmes Rencontres méridionales de Préhistoire récente, Marseille (13): Marges, frontières, transgressions*. Toulouse: Archives d'Écologie Préhistorique, 251–264.

Zilhão, J. 2000. From the Mesolithic to the Neolithic in the Iberian Peninsula. In Price, T.D. (ed.) *Europe's First Farmers*. Cambridge: Cambridge University Press, 144–182.

Zilhão, J. 2001. Radiocarbon evidence for maritime pioneer colonization at the origins of farming in west Mediterranean Europe. *Proceedings of the National Academy of Sciences* 98, 14180–14185.

Index

Note: page number in *italics* refer to figures

Abri des Autours 31, 32, *32*, 124, 138
Actor-Network Theory 1, 77–78, 79–80
Ajdovska Jama 17, *17*
An Corran 66, *96*, 97, 100, 105–107, *106*, 110–111, 142, 191, 203
Ankave eel traps 76, *76*, 92
Annagh Cave 33–34, *34*, 138
Ashberry Windypit *157*, 175–176, *175*, 181, 191
 see also Ryedale Windypits
Ash Tree Cave *157*, 174–175
Ash Tree Shelter *126*, 154
aspect of caves 199–201, *199*, *200*
Aveline's Hole 6

Backwell Cave *157*, 159–161, *160*
Ballymintra Cave 169, 173
Bara funerary practice 48–49
Blue John Cavern *157*, 172, 173
Bob's Cave *96*, 111, 113, 116
Bolóres rock shelter 27
Bower Farm 126, *126*, 142–144, *143*
Broken Cavern *96*, 108, 135, 187, 188, 204

Caisteal nan Gillean II *96*, 97, 99
Can-Pey cave 23, 25–26, 138
Cardial Impressed Ware complex 13, 22, 24, 25
Carding Mill Bay *96*, 100–102, *101*, 110, 135, 203

Carsington Pasture Cave 6, 64
Casa da Moura 28, 134
Cathole Cave *126*, 150, 189
Cattedown Cave 5, 65, *126*, 151–152
Cave Ha 64, 70, *126*, 144–146, *144*, 149, 203
Caverne B, Hastière 31, 146
chaîne opératoire 90–92
Chapel Cave *126*, 146, 210
Charterhouse Warren Farm Swallet *157*, 176–180, *177*, 181, 182, 183
Chasséen culture 24–25, 39
Chauveau CH1 31, 142
Chelm's Combe *126*, 135–138, *136*, *137*, 139, 187–189, 201, 203
Cnoc Coig *96*, 97, 98–99, 110, 135
Covão d'Almeida 28, 134
Cueva de los Murciélagos 24, 134

Darfar Crag Cave *126*, 146–147
Dunald Mill Hole 62
dwelling perspective 73–74, *74*, 79–80

entanglement 78–80, 86, 89, 90, 94, 131, 155, 202

Felsstalle 30, 142
Flint Jack's Cave *157*, 168–169, 201, 203
Fox Hole Cave, Derbyshire 6, 97
Foxhole Cave, Gower 97, *126*, 154
Fussell's Lodge Long Barrow 42–43, 190

Index

George Rock Shelter 65, 86, 87, 96, 111, 113–116, *114*, 135, 138, 187–189, 203, 205–208, *205*
Giggleswick Scar 208–211, *209*
Gop Cave 6
Grapčeva Cave 16–17
Grotta Scaloria 19–20, *20*, 64, 131
Gruta do Cadaval 28, 134
Gruta dos Ossos 28, 134
Grutta I de Longu Fresu 21

habitus 38, 71–73, *72*, 79–80, 84–85
Happaway Cave 6
Hay Wood Cave *126*, 147–149, *148*, 185, 191, 197, 201
Hazelton North chambered tomb 43, 44, 122, 197
Heaning Wood Bone Cave *63*, 109
Höhlenstein-Stadel 29, 124, 138

Ifton Quarry Rock Shelter 6
index (*Art and Agency*) 75–76, 78–79, 86, 88, *88*, 89–91, *91*, 93, 131, 132, 138–139, 181, 184, 202
intermediary period 47–53, 122–123, 125, 131, 132, 135, 138, 139, 145, 149–150, 155, 161, 165, 167–168, 169, 170, 173, 179–181, 182, 184–185, 188, 195–196, 201–202, 204

Jama-Bezdan 59–60, *60*, 168
Jubilee Cave *126*, 139–141, *140*
Jungfernhöhle 29, 138

Kent's Cavern *126*, 154
King Arthur's Cave *126*, 154
Kinsey Cave *96*, 111, 116–117, *117*, 121–122, 200, *201*

L'Abri du Pas-Estret 25, 124, 150
L'Abri Pendimoun 23, *23*
LBK (*Linearbandkeramic*) 13, 28–29
Les Grottes des Barbilloux 25, 124, 149
Lesser Garth Cave 207
Lesser Kelco Cave 126–128, *126*, *127*, 131, 208
Little Hoyle Cave *126*, 151
Little Orme Quarry *126*, 141–142, *141*

Markland Grips 152
Merina funerary practice 50
Michelsberg Middle Neolithic 30–31, 33
Mother Grundy's Parlour *157*, 174
multi-stage burial 2, 4, 18, 21, 28, 34, 40, 43–44, 46–47, 50, 52, 67, 123–124, 125, 150–154
definition 59
mummification 28, 38, 52, 132–134, 185

Nanna's Cave *157*, 161–163, *162*, 203
neotaphonomic research 54–57, *56*, 58, 102
North End Pot *157*, 170–172, *170*, 173, 179, 200, 202, 210

object biography 18, 87–93
Ogof Columendy *157*, 168
Ogof Pant-y-Wennol 6
Ogof-y-Benglog *126*, 154, 203
Ogof-yr-Ychan 6
Orchid Cave *157*, 172, 173, 202
Ossum's Crag Cave *126*, 154

palaeotaphonomic research 54–56, *55*, 102
Parc le Breos Cwm chambered tomb 150–151, 189
pastoralism 15–16, 19, 22, 25, 27 38–39, 122
Penywyrlod chambered tomb 44, 134
Philippine central highland funerary practice 52–53
phreatic environments 61, *61*, 139
Picken's Hole *126*, 153–154, *153*, 201
Pipton chambered tomb 44, 134, 189
Pitton Cliff Cave 6
Pontnewydd Cave 65, *157*, 168
Potter's Cave 97
Prestatyn, Early Neolithic burial 64, 96, 107, 110
primary burial 22, 28, 30, 125, 139–142, 152, 185
definition 57
Priory Farm Cave 6

Raschoille 96, 100, 102–104, 104, 110, 142, 185, 191, 203
Red Fescue Hole 6
Reindeer Cave 157–159, 157, 169, 202
Resplandy Cave 25, 142
Robin Hood's Cave 126, 126, 128–131, 129, 130, 186, 187
Rockmarshall midden 96, 107
Ryedale Windypits 60
  see also Ashberry Windypit

Scabba Wood Rock Shelter 157, 163–165, 164, 169, 204
secondary burial 2, 4, 16, 19–20, 26–27, 29, 31, 34, 38, 42–44, 47, 115–116, 124, 135–139, 151, 185, 187–191, 187
  crania only 121–122, 125–132, 185–187, 186
  definition 57–58
Seine-Oise-Marne Late Neolithic 30–31, 33, 169
Sewell's Cave 96, 111, 119–122, 120, 125, 128, 131, 186, 208
shell middens 98
speleothem 63
  see also tufa
Spurge Hole 111–112, 126, 133–134
Stainton West 73–74, 74
structuration theory 71, 79–80
successive inhumation 2, 4, 9, 15, 24, 25–26, 27, 30, 38, 43–44, 46, 97, 102, 104, 124, 191–196, 192–193, 197, 201–202
  definition 58–59
  Early Neolithic 125, 142–149

Late Neolithic 171–172
Middle Neolithic 156–169

taskscape 74–75
Temple Cave 66
temporality 9, 44–46, 74–75, 79–80, 83, 84–86, 93–94, 173, 184, 185–196, 194, 198, 203–204
Thaw Head Cave 96, 111, 117–119, 118, 121–122, 140–141, 142, 149, 152, 203, 210
Three Holes Cave 108
time categories 81–84
  A-series time 83–84
  B-series time 83–84
Toradja funerary practice 47–48
Tornewton Cave 126, 154
Totty Pot 157, 166–168, 167, 169, 176, 191, 195, 201, 202, 203
Trou des Blaireuax 31–32
Trou du Frontal 31, 32
tufa 63–64, 70, 145, 166, 169, 189, 195, 202–203
  see also speleothem

vadose environments 61–62, 62

Wataita funerary practice 50–52, 51, 187
Wayland's Smithy long barrow 43–44, 45, 46, 190, 191, 194, 195
West Kennet Long Barrow 42–43, 46
Wolf Cave 206–207
Wookey Hole 62

Xaghra 21

EU authorised representative for GPSR:
Easy Access System Europe, Mustamäe tee 50,
10621 Tallinn, Estonia
gpsr.requests@easproject.com